SEVERAL WORLDS

Reminiscences and Reflections of a Chinese-American Physician

SEVERAL WORLDS

Reminiscences and Reflections of a Chinese-American Physician

Monto Ho

University of Pittsburgh, USA

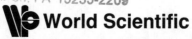 **World Scientific**

NEW JERSEY • LONDON • SINGAPORE • BEIJING • SHANGHAI • HONG KONG • TAIPEI • CHENNAI

Published by

World Scientific Publishing Co. Pte. Ltd.

5 Toh Tuck Link, Singapore 596224

USA office: 27 Warren Street, Suite 401-402, Hackensack, NJ 07601

UK office: 57 Shelton Street, Covent Garden, London WC2H 9HE

Library of Congress Cataloging-in-Publication Data
Ho, Monto.
  Several worlds : reminiscences and reflections of a Chinese-American physician / by
Monto Ho.
    p. cm.
  Includes bibliographical references and index.
  ISBN 981-256-409-8
  1. Ho, Monto. 2. Chinese American physicians--Biography. I. Title.

R154.H578A3 2005
610'.92--dc22
[B]

                                                                    2005050602

British Library Cataloguing-in-Publication Data
A catalogue record for this book is available from the British Library.

Typeset by Stallion Press
Email: enquiries@stallionpress.com

Printed in Singapore by World Scientific Printers (S) Pte Ltd

To
Carol, wife and companion of a life-time,
"Fifty years are like a day."

CONTENTS

FOREWORD

M onto Ho is an outstanding physician, scientist and teacher. He has been a leader in the field of infectious diseases during the past fifty years. He co-discovered interferon, made major contributions to the pathogenesis of virus infections in the immunocompromised host and built a strong, science-based infectious diseases group at the University of Pittsburgh. He could have honorably retired, but wisely decided to pursue a second career in Taiwan. He changed the direction of his research to address problems that were important to that country. He recognized the threat posed by the emergence of antibiotic-resistant bacteria and the need to enhance the quality of training of infectious diseases physicians. He supported able investigators in Tainan and Kaohsiung as well as in the capital Taipei. These efforts paid unexpected dividends. Appropriate use of antibiotics has become an important national health priority. There is now intense research on the devastating outbreaks of enterovirus 71 in children and Klebsiella liver abscesses as well as efforts to improve the use of antibiotics in the community, agriculture and for surgical prophylaxis.

Dr. Ho could not have accomplished his goals without a keen sense of Chinese culture. This required strong alliances and friendships with like-minded leaders such as Drs. Kun-yen Huang and Cheng-wen Wu. He was supported by an advisory committee of distinguished Taiwanese physicians. They welcomed his efforts and protected him from the hostility of a few entrenched adversaries.

This book provides a remarkable opportunity to understand this very special Chinese-born, American physician. It is helpful to learn his personal history, the development of his scholarly qualities and the logic of his scientific and cultural passions. These emerged from the overriding influence of his courageous father, partnership with his gracious wife Carol and education at Harvard and Stanford universities.

The Chinese are delightfully bright people with very human qualities and their own special culture. They appreciate scholars and honor elder statesmen. I had the opportunity to speak about Monto with one of the leaders of the National Taiwan University Medical College, the premier medical school in Taiwan. This distinguished physician was so pleased by the Chinese edition of this book that he ordered 200 copies to inspire his medical students.

One of the greatest pleasures in life is to have collegial conversations with good friends. I had been looking forward to discussions with Monto ever since we met in Boston over forty years ago. He left the Thorndike Laboratory before we had a chance to get to know each other well, but I always considered him to be very special. The time to develop a real friendship came when he invited me to join his teaching program in Taiwan a decade ago. I consider myself privileged to continue to work with him and witness his keen mind, whimsical humor and dedication. This book has answered many of the questions I have always wanted to ask him. It is good to get to know Monto Ho.

Calvin M. Kunin, M.D.
Emeritus Professor of Internal Medicine, Ohio State University
Past President, Society of Infectious Diseases of America

PREFACE

I have just read Monto Ho's Reminiscences and Reflections for the third time and enjoyed it again. I realize that it is about forty years since we first met and I thought that I knew him very well until I read these pages. Perhaps we do not know people as well as we imagine, for I found new Montos in each chapter.

Monto Ho, M.D., is a well known and widely recognized infectious disease specialist, an equally well known virologist, a professor of medicine and pathology, a professor of infectious diseases and microbiology, twice a department head, a chief of division in a medical school, writer of hundreds of scientific and medical papers, husband, father, grandfather, scholar, tennis player, linguist, and, as I now discover, an excellent author. He has lived in the U.S., China, Turkey, Austria, Germany, and Australia and traveled widely. While he obviously loves his Chinese heritage, he is, to me, cosmopolitan, a world citizen.

We first met in 1963 when I visited the department that he eventually led, but I knew about his work several years earlier, when he was doing a fellowship with the pioneer of virology, John Enders. At the time, I was working on anti-viral vaccines and chemotherapy at a commercial research laboratory under the direction of another pioneer in virology, Randall Thompson. From time to time, Thompson would visit Enders' lab. They would discuss problems and exchange materials. On one occasion, he presented Enders with some human cell cultures which I had developed and Monto Ho used one of them in his seminal early work on what we now know as interferon. However, I remember being slightly annoyed that, in his publication, someone had changed the name that I had given to my beloved cell line and it was my intention, when I first visited Pittsburgh, to complain about the matter. I never did. The relationship was immediately so cordial that I lost all motivation to create a scene!

Monto Ho is an introspective person and his reflections are just as interesting as his reminiscences. He has thought deeply about religion, political matters, the nature of the scientific process. In some ways, we have faced the same problems. One knows that his origin is Chinese from his appearance. For me, I have to speak before I am perceived as English. But we are both American and in some ways, typical of the first generation of newcomers; still partly belonging to our original heritage. Read his book. Enjoy it and think about it.

John A. Armstrong, ScD,
Emeritus Professor of Microbiology, University of Pittsburgh

AUTHOR'S PREFACE

This book contains selected autobiographical reminiscences and reflections. Reminiscences are of my background, education and career. Reflections cover a wide field, but together they represent my interests.

Many people believe that one's education is limited to when one was young, or to classroom learning. I feel my education did not stop with my youth or with going to school. It has been continuing throughout my life. I have learned from my elders and teachers, and also from life's experiences. I have learned from books I have read and from discussions and conversations with friends. Reflections have also been part of my education. Reflections consist of the process of reflecting or thinking, and striving for insights and answers to questions. The net result of this type of education has affected my aptitude for problem solving, my *Weltanschauung* and scale of values. It has become part of the better me.

This book also describes and evaluates the main events of my career, which have been in the path of medicine. I understand there is a limit to the extent to which autobiographical writings such as this can be truly objective. Nevertheless, for better or for worse, I have aspired to that ideal in evaluating my accomplishments, whatever they may have been.

In July 2002, I wrote a book in Chinese called *My Education and My Path in Medicine*. It was published by the Publicist Publishing Company in Taiwan[1]. It became known among readers fluent in Chinese, in Taiwan and to some extent in the United States and on the Chinese mainland. This present English version is not a literal translation of that book. It has been rewritten and revised for a wider English-speaking audience, not necessarily familiar with Chinese history or culture. Only 28% of this book is directly translated from the Chinese version. This book has a briefer account of my experiences in Taiwan (Part III, Chapters 13–16). The other chapters are either entirely new, or are expanded from old ones. References of my research work and others and notes are listed numerically under "References and Notes".

I thank Lee Gutkind for help in revising Part I, and Chapters 5, 7–9, 13–15; and teaching me how to write non-fiction literature. Thanks are due to John and Audrey Armstrong for their helpful criticisms and corrections; and Cheong Chean Chian for her careful editing of the manuscript for publication.

Monto Ho, M.D.
Pittsburgh, Pennsylvania

part

I

PERIPATETIC YOUTH

chapter
| 1 |

MY CHILDHOOD (1927–1937)

One day shortly after the New Year, 1936, there was a crisis in the Ho family in Ankara, Turkey. Father discovered that money had disappeared from the inner pocket of his suit coat which was hanging over a chair in the living room. At first, he and Grace, my stepmother, suspected the maid. They quizzed her, but got nothing but denials. Then they started asking me, a nine-year-old boy. I, too, denied knowing anything. They then looked in my drawers and discovered that I had a new fountain pen, which had not been there before. When they asked me where I got the pen, I said I bought it at a store. Since I received no allowance, and had no recourse to cash of any kind, I finally admitted I stole the money. Upon hearing this confession, Father became in turn furious and supplicatory. When furious, he took the poker from the fireplace and beat me violently on my behind and legs. When supplicating, he got on his hands and knees and begged me not to steal again. Stealing is such a horrible crime! His anger and humiliation overwhelmed me. His beating hurt me, as this was first time he had ever used a metal poker. Ordinarily he would use his belt to spank me. But seeing him before me on his hands and knees was devastating. During all this time, Grace stood by without saying a word or moving a muscle.

What surprised me was that in addition to this physical punishment, Father further ruled that I would be shut up in the upstairs of the house for a week, during which I was not allowed even to go downstairs. When my punishment was finally over, I felt dizzy and disoriented after descending from upstairs to downstairs, like I was seasick. This surprised and upset me.

Still, being punished for one's transgressions, even physically, was acceptable. I knew I was wrong. Father did what a "strict father", according

3

to Confucian ethics, was supposed to do. My submission was also part of "filial piety".

My mother had died in Changsha, Hunan when I was five years old. I was in the room as she drew her last breaths. She made a long whining sound which frightened me. They had brought her from the sanatorium to my eldest uncle's house to die. I saw them dress her in a funeral costume. I was numb, partly from grief and partly from fear. My father was far away, getting his Ph.D. in Germany. Still, he was a famous man, being one of perhaps half a dozen Ph.D.'s in the entire province. My uncles arranged an elaborate funeral. Mother's coffin was placed in the living room of my uncle's house for a few days. I was dressed in a rough hemp costume in white, the Chinese color for mourning, and led the funeral procession as the sole surviving son. The procession, unlike Western funerals, was noisy and boisterous. At the head of the procession, there were several individuals with poles on which hung strings of fire crackers that were going off. In addition, a band of musicians playing horns and drums added to the clamor.

Honoring one's mother and father is the major commandment of the Chinese moral code. It is also part of filial piety. Mourning appropriately for the death of a parent is part of being filial. As an unknowing small boy, I had to be helped in fulfilling this obligation — at least I was made to understand its solemnity.

In olden China, Confucian ethics decreed that one mourned for the death of a parent for three years. Officials left their offices for their homesteads and remained in isolation. Even though this rule is no longer observed among the Chinese, the vestiges of the old mourning rites remain. I witnessed Chinese mourning practice while I was in Taiwan. In 2000, I was invited to the memorial service of Ken Wu's father. Ken was president of the National Health Research Institutes (NHRI) of Taiwan, who had recruited me to do medical research in Taiwan after my retirement from the University of Pittsburgh. The service was an elaborate affair that lasted almost a whole day. It took place in a large auditorium, around which there was so much traffic that a corps of policemen were recruited to direct it. The place was packed. The center of activities was a shrine, a tablet on which was written Ken's father's name. On each side of the shrine, amidst an array of flowers, were standing separately about thirty male and female — direct descendant relatives with their spouses — clothed in black. On the walls and scaffolding above the shrine were hung framed calligraphies written for this occasion by notables of Taiwan, including the President and of other highest officials. They were placed in graded positions of prominence, in accordance with the rank of their donors. The main event of the ceremony was groups of individuals; anywhere from a dozen to forty or fifty in number, whether relatives or professional associates of Ken from various social and political organizations in Taiwan; who approached the shrine successively. An official master of ceremonies would announce each group and tell them what to do. They first greeted the relatives standing on both sides of the shrine, and then honored Ken's father by bowing three times reverently in front of it. The number of organizations represented and the social stature of the individuals participating were a measure of Ken's position in society. It took more than a morning for all the groups to pay their respects. I sat quietly among the audience waiting for my turns. I was a member of several groups that went forward. I was a member of the memorial service preparatory committee. This was an honorific committee of about twenty individuals who were socially prominent. I belonged to the Academia Sinica group, and also to the NHRI group.

While waiting, I read Ken's father's biographical sketch, which was in the announcement of the service that I had received. His father was born in Taiwan when it was occupied by Japan. He initially peddled charcoal for a living. Gradually, he covered an area in Taipei full of governmental buildings that included the Japanese governor's mansion. By hard work, he eventually became a wealthy merchant. Although uneducated himself, he was able to send all his children to college, and many of them obtained graduate degrees. Thus I learned the moving story of Ken's father, whom I had never met.

In fact, I would expect that most of Ken's friends and acquaintances in attendance had never met his father. He was a man who died in his eighties. He had been bed-ridden for eight years in Ken's house, after having suffered a stroke. In the West, they would not have attended his memorial service, which would have been considered a private affair. But among the Chinese, this type of service is very much a public affair. Why is that so?

Other questions may be asked. Why is it that among Ken's siblings, it was mostly Ken's associates and friends that attended? I could not get away from the thought that in coming to his father's service, we were really honoring Ken, the most famous of his father's children. Why should the son be honored when the father dies? I cogitated over this while sitting in the audience. It goes back to filial piety. A son's achievements are an expression of his filial piety. The higher and greater they are, the greater the filial piety. And the greater is the honor that can be given to the father.

My mother, Hu Gin-lien, was born in Yiyang. If our family had anything resembling an old homestead, it would be Tau-Hwa Lun in Yiyang, Hunan, China. Yiyang is in northern Hunan, on the shores of the river Tse, which together with the river Yuan flows into the river Hsiang to form a major tributary of Yangtse River. This area is part of the rich farm lands of the Yangtse valley, China's rice bowl. Yiyang is also close to Tungting Lake, one of the largest lakes in China, and full of romance and history. Tau-Hwa Lun was a village across the river Tse from Yiyang, consisting of verdant hills interspersed with rice paddies, but now it is unrecognizably urban and part of the city of greater Yiyang. The name "Tau-Hwa Lun" means "Hills of Peach Blossoms". During the end of the 19th and early part of the 20th century, the Norwegian Lutheran Mission established in Tau-Hwa Lun a campus of western buildings that included a church, a hospital, a primary school and a middle school with separate campuses for girls and boys, a school for the blind and deaf, and an orphanage. During the 1930s, the Swedish Lutheran Mission joined this campus by establishing on it a college called Sin-Yi College. ("Sin-Yi" means "Lutheran" in Chinese.) It was closed down during the Sino-Japanese War (1937–1945). The main building which housed this college was later used by the Sin-Yi senior middle school and it still exists. In 1942–1943, I studied at this senior middle school. The relationship between the Norwegian Lutheran Mission and our family was a close one. I was born in the mission hospital. Yiyang was also the place where my father was born. Father was educated in the primary and middle schools of the mission. He then went to the Yale in China College in Changsha, the capital of Hunan province.

While in college, father married my mother. They were betrothed while both were in high school. Hu Gin-lien attended the Yiyang Lutheran girl's school, which was on the Yiyang side of the river Tse. It was apparent when they were married in Yiyang that she was suffering from tuberculosis, as it was a common disease among young adults, but Father went ahead with the marriage. They kept house in Changsha. Father moved his widowed mother and sister Shau-hua from Yiyang to their new house. Although he and his widowed mother had always been poor, now in college he was relatively well off because he made money teaching English in middle schools in Changsha. He and my mother had a few happy years together. She gave birth to me and then to my sister Man-shia. In the few family pictures that survived which later disappeared during the war, she dressed me and my sister carefully, in expensive western outfits. My mother was a petite, gracious, kind and forgiving person. This is what people tell me about her, but I feel it is true. Shortly after moving from Yiyang to Changsha, my grandmother died. Then my aunt Shau-hua married a provincial civil servant, Wong Tao. She left the household and established one of her own. Throughout their lives, Father, even though he was only three years older, was highly protective of his younger sister, especially after grandmother's death. He considered that part of his filial duty. The "elder brother" is sometimes considered like a father in the Chinese familial scale of values. She worshipped him to her dying day, when she was eighty. One day, Shau-hua went to Father's house and stripped it of all her mother's belongings without warning, and moved them into her own house. My father was furious at her gross covetousness, but my mother placated him, and refused to complain. She said, "Let her have them. They were not that important".

Aunt Shau-hua probably thought that those belongings were hers; she was the beloved youngest child of her mother. Property rights within a Chinese family are fuzzy. The idea is that everyone shares while the family is intact. However, when a daughter marries, she is automatically outside of the family. She belongs to her husband's family. The only way the belongings could be hers was if her mother had explicitly given them to her. Since she was dead, there was no way to determine her wishes. My aunt probably thought that her mother would have given them to her if she were able to. So she just took them. The sad thing is that, had she asked my parents, they probably would have gladly given them to her.

My appearance is overall very much like my father, but the lower part of my face, including my chin resembles my mother. I find myself feeling like her in moments when I am at peace with myself, or when I wish to be at peace with another individual. Both Shao-hua and Father had tempers that could erupt.

A number of calamities shook our family around this time. My younger sister, Man-shia died of diarrhea when she was only two years old when my mother was in the hospital. This was and is a common cause of death in young children in under-developed countries, as China surely was at the time. My father tried to keep the news away from my mother and did not mention her death. One day she told him that she dreamt of Man-shia and she seemed far away. "Is she dead?" she asked. Father had to tell her the truth. Mother could no longer keep house and was confined to a sanatorium at Hsiangya hospital in Changsha for the last year of her life. I went to live with the oldest of her three brothers, my uncle Hu Yi. He taught physical education in Ru Tsai High School in Changsha. His two younger brothers were also frequently in the house.

My mother's funeral procession wound through the streets of Changsha. Finally we arrived at her burial plot in the outskirts of the city. My young mother could not be honored after her funeral like Ken Wu's father. Her son had no accomplishments. Still, she did leave a small son who might yet in the future fulfill his duties of filial piety. That is why they made me lead the funeral procession.

One and half years after my mother's death, my father brought back an American wife to China. He met her in Chicago, where he represented Governor Ho Chien's Hunan province at the Chicago World's Fair of 1933. Grace Lee was a Eurasian, born of a Chinese father and a German-American mother in Chicago. She was a high school graduate and had worked at sundry jobs in Chicago. She was young and beautiful. It was rumored that she had won beauty contests in America, although I had no confirmation of this from my father.

In Father's autobiography which was written in 1990, 55 years after his permanent separation from Grace, he described Grace as "a beautiful woman, with a gentle disposition". This is a 713 page book in Chinese published by the Chinese University of Hong Kong Press. It is called, *A Diplomatic Life of Forty Years*[2]. The book is full of human interest, particularly of people involved in his work. I believe that his description of Grace essentially explains why he married her. I would say that Father was a gay blade in his bachelor days. He was attracted by physical beauty and a gentle disposition in a woman. I know he was not overly interested in intellectual achievements in his wife. I remember before he remarried, he was attracting a number of prospects in Hunan. There was a Miss Liang, who was Hunanese and a graduate Yenching College in Beijing, probably the most prestigious Christian college in China at the time. She visited Hunan one day in 1932 when Father was abroad. There were rumors that she and Father would marry. She visited me and took me shopping. My memory is of her tightly holding my right hand while we were walking in the street. I felt very uncomfortable. I felt the sweat in my hand and I wished to extricate it but I dared not. I did not welcome the prospect of a successor to my mother. Years later, I found out she had become a famous history professor in Taiwan. But we never met again. It is also possible that

Father was attracted to Grace because she was a foreigner. Adulation of the West was part of the culture of China's modern intellectuals. Perhaps he was tired of what was available at home.

Bringing an American woman to live in inland China was a trial, even if valiant efforts were made to ease the way. Hu Ren, my second maternal uncle, took me to Hankow to my father and Grace on their way by boat up the Yangtse River from Shanghai. Hankow was a Westernized international city with foreign settlements about two hundred miles from Changsha. They stayed in a hotel with plumbing and western amenities so there was no problem of acclimatization for Grace there. Before our arrival in Changsha, Father had rented and readied in its suburbs a large two story house in the vicinity of his old alma mater, Yale in China, and engaged three servants to run the house. One of them was Li Ho-sen, who was known for his know-how. It was considered luxurious by local people.

However, there was a crisis the day we arrived in Changsha from Hankow. The servants who prepared the house did not anticipate the fact that westerners are used to sit- down toilets, even if there was no running water. Grace would not stoop down in outhouses like local people. It was apparent after we arrived that she would soon have to "go". She had no way of explaining her problem to the servants, so Father immediately asked Li Ho-sen, to go into town and buy a portable sitting commode. These were available in China for bedrooms of well to do people. They had to be emptied every day in outdoor privies by servants. However capable he was, Li could not have brought back the commode in less than three or four hours. My impression of the afternoon was Grace pacing frantically back and forth in her bedroom, agitated by her urinary urgency while impatiently waiting for the delivery from town. I, a seven-year-old boy who had known her but a few weeks, could not communicate with her either. I could but stand in the corner of the room, watching her speechless and helpless. Fortunately, the crisis evaporated with the arrival of the commode.

After their marriage in Chicago, and before bringing her back to China, Father wrote ahead to my three maternal uncles with whom I was living after my mother's death, that he wanted Grace and me to have a real mother and child relationship. I was to be told by them the story that Grace was my real mother who had been away, and that my deceased mother was just a governess who just took care of me. This they dutifully did. But I was not convinced.

Why did he do this? Confucian ethics do not specifically address the problem of step-parents. Still, like in the West, step-parents are not the same as parents. The filial piety due to a natural parent is not automatically transferred to a step-parent. Father probably thought that as a young child, I needed a "real" mother, not just a step-mother. He was hoping that Grace would play that role and that I would respond as a dutiful son. Perhaps he was hoping this would be possible in view of her "gentle disposition". In addition, he probably felt that as his filial son, I had the obligation to treat his wife as my mother, at least during the time that their marriage lasted; even if I was not convinced about the story of the governess.

One immediate consequence of this fairy tale was that after we arrived from Hankow, all contact with my maternal uncle was cut off by Father. One day, several months later, when I was alone at home with the servants, my second uncle, Hu Ren somehow discovered this fact, surreptitiously arrived in the house and "kidnapped" me for the afternoon. Of the three uncles, Hu Ren was especially fond of me. He was overly affectionate, kissing me and caressing me to the extent that I was at times embarrassed. That day he whisked me away from the house, and took me for a long rickshaw ride in town.

"Monto, your mother died two years ago. Your present mother is not your real mother".

Then he continued with a torrent of verbiage about my mother and her family, the Hu's. He told me that he and my other uncles were my real mother's brothers. This was an attempt to "brainwash" me and disabuse me of the role of Grace. I thought the whole exercise quite unnecessary, as I was quite aware of the facts. My view at the time was that I must be completely loyal to Father's wishes. He wanted me to treat Grace as my real mother. This was what I intended to do, even though I was fully aware that my real mother was dead. I was actually resentful of Hu Ren's behavior, and considered it as shenanigans.

Grace did not know a word of Chinese and I did not understand any English. She would talk to me constantly in English, hoping I would pick it up. A few days after her arrival, I wanted a peach from the dish in the sitting room. At that time any communication with Grace had to be translated by my father, as he was the only other person who could speak English. So I addressed my request to him. He wanted Grace to take charge of such requests, as a mother should. He taught me how to ask Grace in English, which I did with some effort. I asked Grace, "May I have a peach?" copying my father word for word and pointing to the dish of peaches on the table. I eagerly awaited her answer, which I assumed would be either "yes" or "no", words that I had just learned. Mysteriously, she said, "Go ahead". I had no idea what she meant until Father translated these enigmatic words for me. However, progress in learning English was rapid. In a few months, I was able to communicate with her in English.

Grace was not mean. Her attitude toward me was determined by her labile mood, which I found frequently unpredictable. One day she would be gay and solicitous. She would sing her American songs and smile at me. But other days, I felt I did not exist. As we were together much of the time, I would simply follow her around, or play alone in my room. Over time, this detachment was what was hard to take. It was the opposite of love. I tried to overcome this attitude and please her, but in my heart, I felt little affection from her or for her. Two human beings were just co-existing. By trying to predict how she would react, I became good at gauging people's moods and mind sets. I felt I was trained by Grace. She did teach me how to speak English. She taught me the prayer to be said before going to bed, "Now I lay me down to sleep". I always thought the part about possibly dying in one's sleep before awakening somewhat morbid. It actually frightened me.

Life was difficult for Grace in China. The most serious problem was the language barrier. Father was at home in Changsha. He was the prominent secretary general of the provincial government. He had many friends in government and academe. He and his wife were entertained a lot. She should have enjoyed her social status, but she could not feel happy in inland China. To enable her to communicate in Chinese, Father engaged for her a teacher who came once a week. But her progress was pitifully slow. Grace was not good at learning foreign languages. He decided to try for the Chinese foreign service. That was one way for him and Grace to get out of inland China. In 1937 he was offered a job as the second secretary in the founding legation of China to Turkey, which started his career of forty years in the diplomatic service. He was personally recruited by the new ambassador, Ho Ya-tsou, who was a famous general from Hunan in Chiang Kai-shek's army. Father reluctantly left Hunan, where he had worked for Ho Chien, the governor and his mentor, ever since finishing his education.

We lived in Ankara, Turkey, from 1935 to 1937. It was the remote, newly-founded capital city in the center of the inland Anatolian highlands, established by Mustafa Kemal Atatürk (1881–1938), the "Father of the Turks" (Atatürk), a title given by edict of the Turkish parliament in 1933. He was the founder of modern Turkey after the Ottoman Turkish Empire was defeated and dismembered after World War I. He intentionally moved Turkey's capital from Istanbul (originally Constantinople) in order to get away from Istanbul's "decadent" cosmopolitism. When we were there, Ankara literally had only a single modern, paved road. We lived in an apartment house on this road. At the origin of the road, "downtown", there was a square with a huge, heroic bronze statue of Atatürk in military uniform astride a horse on a high pedestal. Living in inland Turkey was difficult not only for Grace but even for my father. No one knew Turkish, furthermore no one was interested in learning it. Ankara was a small provincial town, except for the foreign legations and embassies. The educated Turks and the foreigners spoke mostly French and some German, languages entirely foreign to Grace. There were two movie houses in town. One could visit each one alternately to get all of one's recreation. There were no other amusements. One movie I remember vividly was "Frankenstein". It imbued in me a fear of darkness that remained with me until I was twelve years old. I looked under the bed for Frankenstein or his relatives every night before I went to sleep.

As I grew older, Grace would divert to me more and more household chores. She cooked when the servant was off. I would do all the dishes after our meals, both the washing and drying. One day, when I turned on the hot water heater for a bath, the excess gas around the stove exploded and singed my eyebrows and hair. I was left shaken. When things like that happened, I kept them to myself and did not tell anybody. I was the convenient errand boy. This was especially the case after I became the only one in the family who had learned to speak Turkish. I accomplished this after three or four months in Turkey, by playing with the kids in the neighborhood. Grace would make a list of vegetables and groceries for me to buy in the weekly bazaar of country produce which was held on Wednesdays. At other times, she would take me shopping in the Turkish stores. There I was needed as an interpreter. Grace continued to teach me English. We progressed to the point of her giving me word lists to spell. But her active teaching only lasted a few years. She did not order any books to teach me. I don't recall her reading a single decent book. I don't know whether she lost interest or she didn't have much more to teach. By the time I was a teenager, I had probably exceeded her in English. Her written English was mediocre, and she made frequent spelling and grammatical mistakes.

I had not realized when I lost my mother what the loss really meant. It was only after many years spent with Grace that I began to understand. The love of a mother is the love of a woman; a woman who looks after and fusses over the child's needs, whether meals, clothing, or appearance. A mother is someone who may intervene when a child is severely punished by the father. A mother cares whether or not the child is happy and having fun. A mother looks forward to her child's birthday. A mother sees to it that her child has toys. Even though we could afford it, my childhood after the death of my mother was devoid of toys. A few that I did have I got from my father's friends during holidays, not from my stepmother or Father. And my birthday was never celebrated. There was even some confusion as to when it was. When all these expectations are missing in a family, it is dysfunctional, at least from the point of view of the child. My father might have realized the situation, and he tried to cope with it by having Grace assume the role of my real mother. But it was hopeless.

What I was missing was brought home to me by living with a real family when I was thirteen years old. In the first half of 1939, I attended the American School in Berlin while I lived with an Austrian family, the Prelingers. They had two boys; Kurt was ten and Ernst was fifteen. My age was between them. For the first time, I felt I had two brothers and a proxy mother. The three boys slept in one room and were inseparable companions. Mr. Prelinger was an electrical engineer working with Siemens, the giant German electrical concern, existing to this day. He brought home for his boys scientific toys, like a toy accumulator, a toy electric motor and a generator. They had other marvelous toys that all boys love. They had precision models of warships of the budding German navy in battle gray. The Germans were so good at making attractive toys. There were the battleships, the "Scharnhorst" and "Gneisinau". The famous 52,600 ton "Bismark" had not yet been launched. But there were models of the three famous "pocket battleships", that later became well known to the world. The most famous one was the "Admiral Graf Spee" that fled to Argentina in 1941 or 1942, and was trapped and sunk by three British cruisers as she tried to return to Germany. There was a model of the cruiser "Emden" and models of German war planes, including the huge, four-engined "Condor". I first became familiar with the transoceanic flying capabilities of this gigantic bird through knowledge of this model. The three of us played numerous games, including chess, and I was chagrined that I could not beat little Kurt consistently. Mrs. Prelinger was a fantastic cook; her Viennese family dinners were a daily delight. She would also make my lunch to take with me when I traveled by subway from our house on the Richard Wagner Street to Charlottenburg, where the American School was. Around nine o'clock, we would go to bed. Mrs. Prelinger would go to each of our beds and say good night to us separately. She would sit awhile on each bed and have a little conversation with us. Then she would give each of us a kiss. How I treasured her little kiss! I don't remember being kissed by any of my stepmothers, ever. We lost touch with the Prelingers after World War II and I often wonder what happened to Ernst and Kurt. I fear the worst, for so many perished during the war.

What I did not tell either Father or Grace was the motive for my stealing that day in Ankara. On the New Year's day, 1936, I was walking along the street when I suddenly came upon Ambassador Ho. He lived close by, and was taking a walk in our neighborhood. He knew me well from the parties that he as head of the Chinese mission in Ankara had given. He stopped to say hello to me and wished me a happy new year. Then he dug into his inner coat pocket and took out a fresh ten pound note (a Turkish pound was equivalent to about three American dollars). He gave it to me as a new year present. I thanked him profusely. After going home, I told Grace what happened. She promptly held out her hand to collect my present, saying that she would take care of it for me. I heard nothing more of the money. There were no promises or further plans. The money just disappeared. I felt that I was entitled to the money that I took from Father. I knew I was doing something wrong, because I had a guilty feeling when I went to the store to buy the fountain pen. Still, there was an extenuating circumstance for what I did. At least to me, it was part of the whole scenario. There was no discussion of this incident by Grace. And there was never a private discussion between me and Father about the fact that Grace took the money. She was my mother. I owed her filial piety, which included obedience and loyalty. I did not tell on her. At the end of my punishment, my parents gave me a kite, without an explanation. I was surprised, because they never gave me a gift either before or after this incident. The kite was left untouched because I could not fly it by myself. Was the kite bought with the Ambassador's new year's gift? Was it consolation for the punishment I had gone through? I never knew.

chapter

| 2 |

GOING TO FOREIGN SCHOOLS
(1937–1941)

I t was an overcast day in August, 1938, at the port of Hamburg, Germany. Grace and I had boarded the 22,000 ton liner, "Hansa", of the Hamburg-America Line to sail for New York. We were at the railing of the promenade deck waving good bye to Father, who was on the dock waving back. He had on a long, grey raincoat, and a fedora. He looked so lonely and forlorn amongst the sparse crowd. He would be constantly looking at us, but occasionally he would lower his head, as if in sadness. The ship was casting off. The ship's band struck up with melancholy farewell melodies. I could not withhold my tears.

Our family of three were separating for the first time since Grace's arrival in 1933. She and I were going to the United States while Father remained at this job as the Chinese Consul-General in Vienna.

1938 was a year of crises in Europe. On March 12, Hitler's Germany annexed Austria. I had been a student at the Rudolf Steiner School in Vienna for about two terms (1937–1938), when German troops crossed the border. Within hours, there was a parade of German troops in the "Ring", the main thoroughfare of Vienna. A vast number of jubilant men and women jammed the streets to greet the troops with the Nazi salute and shouts of "Sieg Heil", and "ein Volk, ein Reich, ein Führer" (One people, one nation, one leader). Column after column of smart looking soldiers and vehicles went by. The parade was near our house and I joined the crowd. It was exciting.

Austria being no longer independent, the Chinese Legation was converted to the Chinese Consulate-General, and Father was appointed the consul-general by the Chinese government. A plebiscite was held a few weeks later and 99.7% of the vote was for "Anschluss" (merger or annexation). But this was not the vote of thousands of Jewish citizens of Austria. An immediate pall fell over them, as they knew that the Nazis would institutionalize anti-Semitism. While school continued for the term, great uncertainty ruled as there were many Jewish instructors and students in our school. Toward the end of the term, we invited Dr. Hiebel, my teacher at the school, for lunch and to say goodbye. He was a Jew and he was considering the melancholy prospect of leaving his beloved native Austria. In the fall of 1938, the school was closed when Grace and I were in the United States.

In the summer of 1938, Hitler precipitated a crisis in Europe by turning on Czechoslovakia after annexing Austria. As a starter, he demanded the Sudetenland, an ethnically German western rim of the country, that had been disposed of by the treaty of Versailles after World War I. Hitler's threat went beyond the Sudetenland, which would have been gladly ceded to him by the Allies, the British and the French. He challenged the very existence of Czechoslovakia itself, which only had the haphazard support of Great Britain and France. Hitler's increasing belligerence was bringing Europe to the brink of another world war. Under these circumstances, Father decided that for safety, Grace and I would go to the United States. That was home for Grace, and I could easily go to school there.

I attended the sixth grade at Public School Eight in Brooklyn, New York in the fall of 1938. By November, the Czech crisis had subsided following the famous appeasement of Hitler by the British prime minister, Chamberlain. Not only was the Sudetenland ceded to Germany, eventually the entire country of Czechoslovakia ceased to exist. The Czech part became German Bohemia and Moravia, and Slovakia became a German satellite state. In the spring of 1939, Father thought it was safe enough for us to return to Europe. To continue my education in an American school since my school in Vienna had closed, my father sent me to the American school in Berlin where I stayed for a term with the Prelingers.

During the four years of political upheaval between 1937 and 1941, I attended five different foreign schools, two in Europe and three in the United States. I did not attend Turkish schools during the three years we were in Turkey as Father tutored me at home (1935–1937).

We were still in Vienna in September 1939 when World War II broke out. Hitler turned on Poland as he had the previous year on Czechoslovakia, but this time Great Britain and France stood behind Poland and refused to cave in. They declared war on Germany following Germany's invasion of Poland. Food rationing took place immediately. Whole milk was no longer available, only "Magermilch" or "skim milk". Air raid precautions were taken and nights were totally blacked out. But for us, life was pretty normal. This was the lull before the real fighting of World War II. In April of 1940, the three of us left Vienna and Austria by automobile. Our driver, Rudolf, who had been with us all the years we were in Vienna, drove us through Switzerland and south to Trieste, Italy where we boarded the ship "Saturnia" for the United States.

Going to so many different schools in such a short time was not so much a problem of catching up, but rather of adaptation to new environments. With each different school, I had to make new friends. What I dreaded most were the first few days or weeks of going to a strange classroom and meeting the stares of unfamiliar classmates. The loneliness and isolation was most uncomfortable. If one can overlook these discomforts, there is a great deal to be said for going to different schools in different environments. The other major challenge was having to deal with foreign languages.

Coming from the Chinese Legation in Turkey in 1937, Father was transferred to the Chinese Legation in Vienna, Austria, where he was promoted to first secretary. This change involved our moving from a rather remote out-of-the-way capital to the previous romantic capital of the Austrian-Hungarian Empire. Although Vienna was no longer the capital of an Empire, it was still the capital of Austria, and one of the most important cities in central Europe. It was a center of culture and arts. The other big change was the difference in languages. German was spoken in Vienna. Father had studied in Germany, and he spoke and wrote German fluently. This was quite different from the situation in Turkey, where the native language was shunned by most foreigners. German was an important international language at the time, especially in the sciences. The three languages considered internationally important were English, German and French. Father was eager for me to take up the study of German, formally and seriously. We arrived around June and Father immediately engaged an Austrian girl to tutor me in spoken German in order to prepare me for entrance in the fall into a German school. Within two or three months I was able to speak and understand ordinary German.

In the fall, I started the fourth grade of a German-speaking school in Vienna, the Rudolf Steiner Schule, which is related to the Waldorf schools also known in the United States. Years later in 1965, Bettie, our daughter, attended the Waldorf Schule in Tübingen, Germany while we were there for my sabbatical from the University of Pittsburgh. She was also placed in the fourth grade after being in Germany for only three months. We were both able to keep up with the school work, as we had learned to comprehend the language and express ourselves in rudimentary German. Naturally, we had to work to catch up with reading and writing German in class.

Because of the unusual circumstances of my childhood, I was able to achieve extraordinarily in learning languages. Before the age of eleven, I could speak four different languages: Chinese, English, Turkish and German. Having to live with Grace, my stepmother, forced me to learn English quickly in order to survive in the family. This experience no doubt acclimatized me to learning foreign languages. Turkish and German followed in its wake. This experience has given me certain insights about learning a foreign language. The learning of language is part of human nature. One's ability to learn a language, however, varies with age. One is very proficient during childhood but one becomes less proficient after adolescence. The scientific explanation of this is not entirely clear to me. But according to my observation, before the age of thirteen, it is possible to learn a foreign language with ease and fluency given the proper environment. One can pick up the foreign language and acquire a native accent. That is, one's native tongue does not interfere with one's pronunciation of the foreign language. After that age it becomes more difficult. One's ability to advance in a spoken foreign language is greatly facilitated by acquisition of a proper accent and pronunciation. Every language has its natural syntax, rhythm and inflections. The acquisition of this, which includes the grammar of the language, is simpler when one is young and if one acquires the proper pronunciation.

That age thirteen is a cutoff point comes from my own experience and my observation of the former U.S. Secretary of State — Henry Kissinger. Kissinger came to the United States at the age of thirteen as an immigrant from Germany. Although he learned to speak and write flawless English, indeed, superior to most native speakers, his spoken English remained larded with a heavy German accent. The four languages I learned to speak well were all learned before I was eleven years old. I could speak them without an accent. My stepmother was in the same countries as I was, and she had the same opportunities to learn Chinese, Turkish and German. But she couldn't learn any one of these languages well. Although she was not yet thirty years old, because of her imperfect pronunciation, her progress in learning those languages was painfully slow.

However, while small children learn foreign languages quickly, they also forget what they learn quickly. After I left Turkey I had completely forgotten all my Turkish within six months. I literally retained not more than half a dozen words thereafter! Interestingly, my son John had the same experience sixty years later at about the same age. He had been fluent in German after being in Germany for ten months. German vanished from his memory after we left Germany.

Those who knew less Turkish than I did, such as my father and Grace, retained much more for a longer period of time. I learned that this phenomenon can also be corrected with effort. I was in German speaking countries only as long as I was in Turkey, that is three years, but I made a particular effort to maintain my German after I left Germany. A language can only be kept up by constantly reviewing the language. I made it a habit to review German in my mind from time to time in order to keep it up. That is, I practiced speaking and thinking in the language by myself. I also took every opportunity to read and write and broaden my knowledge of the language, even when I was in remote parts of war-time China, and during my long time in the United States. Facility in reading is essential for keeping up a language. As a consequence, I am able to maintain proficiency in the language, even though, inevitably, much has been forgotten. The reward of my persistence in maintaining German was that during my sabbatical in Tübingen, Germany in 1965, I was able to reach the point where I gave scientific medical lectures in German.

Knowing one foreign language well helps the learning of another. We often see people from eastern Europe, particularly Slavic people, such as Russians, Poles, Serbs, who can speak three or four or even six or seven different foreign languages with relative ease. Of course, the types of foreign languages they usually learn are all interrelated. That is, they are Indo-European languages either of Germanic, Latin or Slavic origin. But if one succeeds in learning one foreign language well, it facilitates the learning of another. Proficiency and learning becomes a habit. This is in contrast to the United States where children are usually taught foreign languages in classrooms for only one or two years. A smattering of Spanish, German or French learned for a short period of time is almost useless for mastery of a foreign language. That is the reason why ordinary Americans do not speak or know foreign languages proficiently, even though they may have tried to learn more than one. More effort and time must be exerted. Besides the four mentioned languages, after I was twenty years old, I also studied French and Russian. Of course, my proficiency in these two languages is not as good as my Chinese, English or German because I didn't learn these languages in the native environment where they are spoken, and I was much older. I learned French in Chinese and American colleges in 1947–1948 for a total of two years. This was not a very long time but I was very interested in the language, studied hard and enjoyed learning it immensely. What's more, I kept it up. Such was my proficiency that in 1965 when our family traveled through France for the first time while we were going to Germany, I was able to speak and converse in rudimentary French and make myself understood. My wife, Carol, and my children were amazed. My explanation of my achievement is that the learning of French was helped by my prior combined knowledge of English and German, two related languages, not only in syntax and grammar but more importantly in pronunciation. For example, the characteristic accent of English speaking peoples when they speak French can be corrected if they know German well. The classmates in China and the United States with whom I learned French did not have this advantage.

The problems of learning foreign languages are also brought out by reviewing the history of the Chinese learning English. English has a history of at least 150 years in China. During this time, English became the first foreign language. In my father's younger days, learning English was of primary importance in his education. He frequently cited English, Chinese and mathematics as the three most important subjects of my own secondary school education. Even under communist China, which was violently anti-American in the 1950s and 1960s, English did not entirely disappear from the curriculum, although it favored Russian for a short period of time. It is now unquestionably again the first foreign language.

For over fifty years following the end of World War II, Taiwan has had an uninterruptedly intimate relationship with the United States. All high schools teach at least six years of English. Those who go to college add four more years. Therefore, the average college graduate has ten years of experience learning English. But my observation is that this gargantuan effort is not worthwhile. After spending ten years studying English, the average college graduate cannot speak English, cannot understand common English conversation, cannot read ordinary English books and journals, and cannot write a paragraph of English without grammatical mistakes. If among these four skills of handling a foreign language one cannot achieve any one of them, is this not a waste of time?

If we observe how a child learns a language, the basic skills of understanding and speaking are learned by being in an environment where the language is constantly heard and spoken. The American armed forces developed during World War II an effective method of teaching foreign languages by using this insight. By a method of total immersion, they were able to teach any G.I. to understand and speak perfectly any foreign language in six months. I remember meeting some veterans who were my schoolmates at Harvard College in 1948, who had undergone such training. Some of them addressed me in perfect Chinese, without accent or mistakes. They even learned how to curse fluently. The products of this method are a striking contrast to the graduates of Taiwan's and China's schools.

It is clear then that success in learning a foreign language is highly dependent on the methods by which it is taught. Learning a language is like learning how to swim. Both are acquired skills. After a certain point, one either knows how to swim or one does not. What is the method of teaching English in Taiwan? Taiwan is a highly literate society where doing well in school is almost the primary mission of growing up. Passing examinations and getting ahead are major preoccupations in childhood. I would frequently look at the exam questions in review books sold in ordinary bookstores. The questions about English were almost uniformly on details of rules of grammar, single sentence construction, and vocabulary. In Taiwan and on the Chinese mainland, the teaching of English grammar has become almost an obsession. I feel this is a big mistake. Because after ten years of studying English grammar, the student is still unable to say or write a few sentences in English without grammatical mistakes. The fact of the matter is, concentrating on grammar cannot assure correct usage of the language. Again, we must learn by observing the child. Children speak grammatically correctly without learning any rules of grammar. In order to speak and write correct English, one has to get into the basic rhythm, tonal qualities, and the spirit of the language. One must be exposed to and immersed in speaking, listening to, reading and writing the foreign language. One day I read a comment in the local English newspaper in Taipei by an American teacher of English. He said that in Taiwan the teachers of English do not "teach English" but "teach about English." Interestingly, I believe the same comments may be made about learning English in Japan's schools.

Another relatively ineffective way of learning a foreign language is to recite word lists. I remember my father telling me about this when he was a student. One morning, more than fifty years later, I observed in Taiwan one of our student employees doing exactly that. She was poring over a dictionary. She thought that was a good method to learn English. The fact is, most people do not improve their vocabulary by studying word lists or the dictionary. Native speakers of English or Chinese increase their vocabulary by speaking, listening to, reading, or writing the language. Studying grammatical rules, concentrating on the structure of sentences, and reciting vocabulary lists are all ineffective by themselves. All these are subsidiary methods but cannot be used as primary methods. If one keeps on doing this year in year out, the results are useless. It is like learning how to swim without getting into the water.

The Chinese and Japanese languages are linguistically totally different from English and other European languages. Not only is it difficult for a Chinese person to learn a European language, one also observes that a European has tremendous difficulty learning Chinese. A German or a Northern European, such as the Dutch or people from the Scandinavian countries, learns English with greatest ease. This is because languages in these countries and English all belong to the same Germanic stock. Southern Europeans, such as the French, Italians or Spaniards, have slightly more difficulty in learning English, because their languages, though related to English, are further removed.

One can observe the difficulties the Chinese have learning English by observing immigrants from China in the United States. I noticed that when such a person with a background of school English in China comes to the United States, he or she will improve rapidly for about five to seven years. After that, they reach a stage where no further improvement is evident. Therefore, one observes that a Chinese person even after twenty or thirty years of residence in the United States, still cannot make a speech without grammatical errors or write a page of English without common mistakes. The reason for this is that a solid foundation was not laid in the beginning. Just like in swimming, the student has learned to swim, but his style is faulty and can no longer be corrected. The types of grammatical mistakes which a Chinese person makes are also typical. Very often, it is in the use of the singular or plural, articles and prepositions. Interestingly, a German immigrant, even if he speaks with a thick guttural accent, will not make such mistakes. We also observe situations in places where use of English is commonplace. In Hong Kong and Singapore, most of the educated people learn to speak English fluently. But fluency does not necessarily mean perfection. Very often, a different form of English related to pidgin English may emerge. These are some of the problems of learning a distantly related foreign language after adolescence and without optimum instruction.

Everyone agrees that English has become so important as a global language that learning it has become a necessity. Since it is so difficult for an adult Chinese person to learn English perfectly, it is incumbent on them to reassess their goals. Among the four different language skills, that is, speaking, comprehension, reading and writing, which skill is the most important one? It seems to me that among the four different skills, the easiest one to achieve and the most important one is reading. It turns out that among the teaching of the four different skills, this particular one has been the most ignored in Taiwan. The purpose of learning a foreign language is to extend one's intellectual horizon. Whether in commerce or science, one must be able to access the literature in that language. If we can go beyond one's occupational interest and extend the access to appreciation of books and periodicals in English, then all for the better. It is a shame that after ten years of learning English, the college graduate in Taiwan does not avail himself of these treasures.

Even in the United States, the "three R's" — reading, writing, and arithmetic — are considered the basic skills. Reading broadly and intensely is emphasized in the United States much more than in Taiwan or in China. The tendency there is to read intensively but narrowly. Emphasis was placed in the old days on reciting just a few books. This traditional habit has to change. Recently in Taiwan, a new emphasis has been placed on even reading broadly in Chinese. The same practice should apply to English.

As for learning comprehension and speaking in English, one has to go back to the experience of children and the immersion method. Only a proper teacher, preferably one whose native language is English, can provide the proper environment for such learning. This of course is one of the great difficulties when one thinks about teaching English on a mass scale. Such teachers are simply not widely available. Failing that, I would reorient the teaching of English to teaching reading.

My experience with languages has been important in my life. Without proficiency in Chinese I would not have understood Chinese culture or remained attached to it throughout my life. I would not have done medical research in Taiwan in my late years. Proficiency in English helped me to become a seasoned American. To be able to speak and write English well was essential for my medical career and for my research in America. German, French and rudimentary Russian have opened my eyes to cultures where these languages are native. Reading in German and French has become a lifelong habit. These languages make me feel like an international person. I feel enriched because of them.

The Rudolf Steiner school I went to was situated in Graben, in the middle of downtown Vienna, near the landmark Saint Stephen's Cathedral in the first district. We lived at 3 Beethoven Place in the third district, across the street from the Stadt Park, or City Park. It was a brisk half hour walk through the main streets of Vienna from home to school, which I took every school day. The school emphasized different forms of expression, not just the usual subjects taught in school. There was more than the usual attention paid to art and music. There were special activities such as "Eurhythmy", or graceful exercises, sewing and knitting, and cooking.

Our German teacher, Mr. Hiebel, taught us Goethe's dramatic poem, "Erlkönig" or "King of the Elves". It describes the midnight horseback ride of a father holding his frightened child through haunted woods. Mr. Hiebel read the melodious, guttural German words with flamboyant drama and graphic gestures. The child described his vision of the crowned elf king with his scepter, tempting the child to come to him. The king's dancing daughter beckoned him. The father said that what he saw was a streak of fog, and what he heard was only the breeze. Finally, the boy said the king said, if you do not come, then I will take you by force. Coming out of the woods and arriving at an inn, the child was found dead.

Talking about Goethe, one of my father's friends in Vienna, Frau Burger, invited me to stay with her for a month in the summer vacation of 1939. She was the widow of a professor at the University. Both she and her sister, whose husband was a movie producer, were close friends of Father's. She and I were alone in her elegant house in the delightful suburb of Grinzing. A servant cooked all our meals. We would dine together for three meals a day, in style. I emulated her elegant table manners. The month was devoted to study and cultural amusement. For study she tutored me in a course in German literature. She concentrated on Goethe. To tell about his life, she had me read a number of booklets suitable for children. I learned about his incredibly eventful life of 82 years. He had so many girlfriends! One of them inspired his first bestseller, "Werther", which took Europe by storm. Werther committed suicide because of his hopeless love for a married friend. He wrote it when he was about twenty years of age. Then he wrote so many poems, which are understandable to German reading people of all ages. She did not introduce to me his definitive epic drama, "Faust". I read this later in my life. Goethe was a genius for all ages and for all times.

Frau Burger also introduced to me the dramatic arts. She took me to the classical state theater to see Schiller's play, "Maria Stuart". She took me to see Verdi's "The Masked Ball" at the Vienna State Opera House, which was later destroyed by air bombardment and restored after World War II. I saw this opera for the second time in Pittsburgh in 2003. Interestingly, the play by Schiller was also performed in Pittsburgh around the same time. In order to understand these dramas, I listened to her telling the stories and I had to read diligently. I struggled with Schiller's play in verse.

Then she gave me a project which took me a month to finish. This was to draw a map, to scale, of Grinzing. I measured the length of the streets by counting the steps by walking. Frau Burger was very kind and very enthusiastic. Her influence on me was profound. What she did with and for me were such pleasant surprises. She also introduced to me aspects of Western culture outside of Father's interest or competence, such as German literature, music and practical science.

Another teacher at the Rudolf Steiner Schule, was Mr. Zwiebach. He enraptured the class by reading in German to us from the famous Chinese novel, The Monkey King or The Western Travels, a story of the surrealistic adventures of a Buddhist monk and his monkey-faced assistant, traveling from China to India to obtain Buddhist scriptures in the Tang dynasty. The arduous trip took not just months, but years. It was only later that I became acquainted with the novel in Chinese. It is a parody of a famous, real monk of the time, Xuen Chung.

At that time in Europe, Latin was an important subject in all schools. Even though I only learned Latin for a few months in the second semester, I was attracted by its grammatical complexities; six cases! Even German had only four. The pronunciation of Latin words, entirely phonetic but different in Germanic countries than in Anglo-Saxon countries, has remained with me with the Latin sayings that Mr. Hiebel, who also taught us Latin, gave us to recite. Latin is the origin of many European languages. At one time, it was the universal language used in the western world. I have since then always eagerly read, but not necessarily comprehended the Latin in the many European monuments I have visited throughout my life.

I made some good friends in the class. Karl Hauswirth lived near our house and we walked to school together. I went frequently to his house to play. His parents were first cousins, and he explained that was the reason that his older brother was born deaf. Fifty years later, during a visit to

Vienna, I decided to look him up in the phone book. And I found him! We spent an afternoon reminiscing about our childhood. Our school was liberal and easy going. In the beginning, I was regarded with curiosity, largely because of my appearance. Most of my schoolmates had never seen an Asian or Chinese before. But before long, I felt I was one of the group. I remember being unhappy because I did not feel the school had enough work in the "tough" subjects, such as mathematics and science. Later, I found out that some Germans did not believe Waldorf schools prepared a student for the university as well as the regular "gymnasiums" (preparatory high schools). But for me and for my daughter Bettie, we were in primary schools and we had a fantastic introduction to German and German culture.

Three of the five foreign schools I attended were located in Brooklyn, New York. All together I stayed in these American schools for two and half years, so they made more lasting impressions than the schools in Europe. The first American school I went to was Public School Eight in 1938. After returning to the United States from Europe in 1940, I attended Public School Nine for one year. I joined the graduating class of the primary school, the eighth grade. These public schools in New York City were ordinary, plebian schools without any pretense or sophistication. They were schools in middle class and lower middle class neighborhoods. My classmates came from families with diverse socioeconomic and ethnic backgrounds. There were some blacks and many children of immigrants.

America is avowedly "a melting pot" of many peoples. People of very diverse origins, immigrants, and especially children of immigrants, rapidly become assimilated, partly by attending schools like the ones I attended. The American culture has a powerful assimilating effect. One example is the rapidity with which the children of immigrants forget or want to forget their native languages. They want to learn "American" and be rid of any foreign accents. I believe that one reason why children do not learn foreign languages well in American schools is that inadvertently there is discrimination against all foreign languages, as one of the first tasks of becoming Americanized is to maintain the primacy of the English language. One proof of Americanization of immigrants or their children is the rapidity with which they forget their native language. An example were the children of Austrian Jewish immigrants I had known in Vienna. They resolutely refused to speak with me in German, even when I addressed them in that language. Of course, they had more than one legitimate reason to forget German. More than sixty years later, American prejudice against foreign languages has diminished somewhat, as Americans appreciate more the importance of maintaining their diversity.

American culture emphasizes making money, in getting ahead in one's business, work or profession. America is a land of so many opportunities that "getting ahead" does not necessarily require much formal education. This is true of the glamorous professions such as professional sports and show business, which are particularly enticing to young people. Some professions require a lot of education. But in the primary schools I went to, preparation for the professions was not the mainstream. Most graduates did not go on to college. The ordinary American does not place as much emphasis on intellectual achievement, as I was trained to do as a Chinese. It is not part of the work ethic, and students do not study as hard. This is the reason why a Chinese or European child upon attending an American school not infrequently excels in academic subjects.

One of my typical "pals" at P.S. 9 was Bob Jones, who lived two streets from our apartment at 361 Sterling Place. He was a lively boy with red hair who was about my size, on the small side of medium. We lived near Grand Army Plaza near Prospect Park in Brooklyn. We went to school together. I went to his house after school and on holidays almost every day. Bob's father was a foreman, and his mother also worked. He had a younger brother. We played games and listened to the radio together. We rooted for our favorite baseball teams, which in Brooklyn were usually the incomparable Dodgers. Bob, however, rooted for Pittsburgh and the Pirates, as this was his father's hometown. Bob was extremely proficient in a softball-like game called "stick ball", which was played in the streets using a soft rubber ball and a broom stick. This is a very exciting game, because if the ball is hit well it can go very far, much farther than a softball. Most of the kids, even when not playing the game, had a rubber ball in their hand and would bounce it off walls and the outside stairs for practice. I was one of them. Bob belonged to a team which went to different streets to play other teams. I was envious of him, as I was not good enough to be on the team. I watched him. I excused myself by believing I had not been long enough in America to develop the expertise. But I sometimes think that I may never have caught up. I may not have been adequately athletic, or maybe like reciting classical Chinese, it is almost like something one is born with. It was part of the American emphasis on sports.

I did well in the two public schools that I attended in Brooklyn. Before graduating from P.S. 9, I was recommended by the school to take the entrance examination to Brooklyn Technical High School. There is a system of superior "regents" high schools in New York City which accept students by merit and are not restricted geographically. I was one of two students from our class who were recommended to take the examination. The other student, whose name was Constantine, was son of a Greek immigrant. He was a clean-cut boy, with straight brown hair. On the morning of the examination, I went to his house to meet him. I had to wait downstairs because he was busy praying for success on the examination in his room. We both then took the subway to get to Brooklyn Tech for the examination. I was the only one who passed. I felt sorry for Constantine, because he was one of the students in our class who was conscientious and studied hard.

Upon attending Brooklyn Tech, I felt immediately that this school had a higher academic standard than the public schools that I had attended. The difference in mathematics was particularly striking. Brooklyn Tech had two tracks. One was a terminal track intended for students who wanted to go to work after graduation. The other was for students who wished to go on to college. The one college that Brooklyn Tech prided itself of preparing for was MIT in Boston. Vaguely I thought I was heading for this second track, even though I left before the decision had to be made.

During my first term at Brooklyn Tech in 1940, every morning for a period of two hours there was a compulsory subject called "shop", during which various types of wood work and metal work as well as tools were introduced. One morning, our Scottish instructor, Mr. Kirkwood, was giving a long lecture about safety in shop. I was feeling bored and started playing with the tools sitting at my desk. After class Mr. Kirkwood came to me and told me that he noticed what I was doing and that it was wrong. He punished me by making me write a composition entitled, "Why I should not play in Shop". He gave me a week to complete the assignment. After a week, I handed in my assignment and heard nothing more about it until the end of the term. I learned then from my English teacher, Miss Bennett, that Mr. Kirkwood had come to her with my composition, showed it to her and, suggested that on the basis of that composition alone I should be given a grade of "A" in English. But he still only gave me a "C" in shop. This incident suggests that I was better at expository writing than at shop. It is probably a good thing that I had to leave Brooklyn Tech after a year. I was not inclined manually and probably was not suited for an engineering career. Engineering might not have given me an outlet for the type of problem solving that I became good at. If I had felt obligated to continue in engineering, as the traditional discipline emphasized in my training might have inclined me, my optimum development could have been compromised.

chapter

| 3 |

RETURN TO CHINA (1941–1947)

"You have arrived just in time for tonight's party," said my father as he and I embraced upon my arrival at his house in Chungking from Yiyang, Hunan, in January, 1943. "But first you have to take a bath, and then we have to get a suit for you to wear."

I was then almost sixteen. We had not seen each other for a year and half. We last said good bye to each other in Hong Kong, in September, 1941 when he had flown to Chungking, and I was left in the ninth grade in a boarding school after we both had come from Brooklyn, N.Y. I was stranded in Hong Kong after the Japanese bombed Pearl Harbor and conquered Hong Kong in December. I escaped Hong Kong in early 1942, and had been going to schools and hopping around inland China in Lien Hsien, Guangdong and in Yiyang, Hunan, since.

During this time, my father's private life had also undergone a major change. But this was for the better! In March, 1942, he married my second stepmother, Huang Shau-yun. I met her now for the first time.

"You must be very tired. There is time for you to take a nap before we go to the party," she said.

She instructed the servant to ready a bath for me. I was assigned to my own room in the cute little brick bungalow that they had just built. They had returned but a few months ago from the United States, where Father was counselor of the Chinese military commission.

A wooden tub was placed in my room, and the servant fetched buckets of hot water for me to bathe in. I basked luxuriously in the warm water, going over the unbelievable events I had gone through since I last saw Father.

I could not believe that I had finally come home. I continued to think as I tried to take a nap. But I could not fall asleep.

After my nap, my stepmother handed me a suit of my father's to try on for size. We were now almost the same size, although he was still a bit heavier. I was able to wear the suit. It was a dramatic change from the student uniform that I came in. It brought me back to the days in the United States! The only difference was that I had a military crew cut.

After I got dressed, Father and I talked. He was so happy, grinning and laughing off and on. There was so much to talk about! We started in Chinese, then switched to English, and ended up speaking in German. We both enjoyed foreign languages. They reminded us of our seven years abroad.

That night we attended a dinner and dancing party. Most of the people were Father's colleagues in the Ministry of Foreign Affairs and their spouses. They all knew about me. They came to speak with me, one by one. How my father suffered in anguish when I was incommunicado in Hong Kong! No news for three months! They commiserated with him. And then, almost just before the day of his wedding in March, 1942, he received a telegram from me in Ku Kong, Guangdong, in free China. I had just safely arrived from Hong Kong. They rejoiced with him, celebrating his two blissful events.

After dinner, I joined in the dancing. I danced with my stepmother. She made a very favorable impression on me. She was considerate, thoughtful and seemed really pleased to see me. Although we had never met, we had gotten to know each other by corresponding with each other after their marriage, one letter a week.

Huang Shau-yun was different from Grace Lee, my American stepmother. She was a native of Hunan, Father having taught her English in middle school while he was a student at Yale in China College. She, too, was an extremely beautiful woman. But she was at home among the Chinese and she was naturally gifted in dealing with people. She was happy wherever Father was stationed. My relationship with her was different from my relationship to Grace. There was no fiction about who she was. Although we never lived together for any long time, I felt she was concerned about my health and well being, and we corresponded faithfully when we were apart. When I was in college, she ordered for me a long and warm Chinese gown for the winter in Beijing, which I always remembered. With her, I no longer expected the type of maternal love that I got to know about with the Prelingers in Berlin. Perhaps our greatest bond was our concern for Father.

He was our common object of devotion. He was almost our sole topic of conversation when we were together. I liked this because I learned details about his daily life that I would otherwise be unaware of. I thanked her many times in their later life together for her care of Father. His life would not have been happy without her. Their happiness was complete when she gave birth to a daughter, Manli, in 1951, nine years after their marriage. Their married life together of fifty-five years was a blessing.

How did this series of adventures start? How did I get to Hong Kong?

One day in America, in the spring of 1941, my father asked me, "What would you rather do, go back with me to China to study, or remain in the United States? If you remain here, you can continue to go to Brooklyn Tech, and I think I can arrange with the Walters downstairs for you to stay with them. If you come with me back to China, you can attend a Chinese school. But there is the risk of the war."

At the time, I was a happy first year high school student at Brooklyn Tech in Brooklyn, N.Y. Father had just received orders to be transferred back from New York to the war time capital of China, Chungking. I knew this. What surprised me was that I, at age fourteen, was being asked to decide on my own future.

We had been abroad now for seven years (1935–1941), a good part of my childhood. One of my father's preoccupations during this time was to keep up and advance my Chinese. He had brought with him a trunk load of Chinese textbooks, covering all subjects from primary school to middle school, which accompanied us everywhere we went. The books on Chinese language were the ones he made most use of. He would give me a Chinese lesson three or four times a week. The way the lesson went was like this: he would assign a lesson from the textbook to me. I would go over the lesson and note the characters that were unfamiliar. Then we went over the lesson together. He gave me the pronunciation of the characters I did not know in Hunanese, a Chinese dialect close to Mandarin. As a mnemonic device, I wrote at the side of the character a homonym or another character with the same sound. I would then study the lesson by myself and he would test me on that. For writing, he had me write a diary in Chinese. In this way I went through textbook after textbook, roughly in accordance with my proper grade.

The Chinese language is unique in many ways. It is the vehicle of Chinese culture, a great culture, one of the oldest and a viable one. I have often felt the strength of Chinese culture throughout my life time. The Chinese classics, like the analectics of Confucius, were written 2,500 years ago. Yet they are understandable as if they were written yesterday. The same is true of Chinese poetry. There is no comparable analogy in other cultures of the world. The reason is that written Chinese is not a phonetic language, like most other great languages. It consists of pictorial ideograms. Ideograms give what is written in the language permanence, and a unique capacity for wide distribution. Japan, Korea and Vietnam adopted written Chinese without necessarily learning to speak Chinese. Modern written Japanese remains partly Chinese. Phonetically based written languages are more transitory. We can barely understand the English of Chaucer, and he lived less than 600 years ago. You can read about and study Chinese culture in other languages, but because it is available in its own language, to really understand and appreciate it, requires the knowledge of spoken and written Chinese.

Besides learning Chinese, another aspect of being Chinese, particularly during the Sino-Japanese War (1937–1945), was that one was patriotic and wanted to save China. For my father and me, even before I was ten years old, to be patriotic was probably the most abiding ideal we had. It was much more important than any other ideal, religious or otherwise. It is difficult nowadays, when the United States is not even a friend of China, to understand the depth of this passion. Chinese not only honored their country, but had a heartfelt compassion for China, which was the fuel for this abiding patriotism. The compassion was brought about by China's long history of humiliation by the western powers and Japan. It began more than a hundred and fifty years ago, with the defeat of the Ching Dynasty by the British in the Opium War of 1841. Finally, beginning with the Sino-Japanese War in 1937, the Chinese were at least fighting back, and were not defeated. The Chinese had something to stand for and were standing tall.

In thinking over the alternatives my father gave me, I felt that even with his teaching me Chinese, as years went by, it was getting more and more difficult to keep up Chinese, let alone advance it. Because of the pressure of my schoolwork at Brooklyn Tech, the number of lessons he was giving me had gradually decreased to one a week during the school year. I felt I was constantly losing ground in the study of Chinese in the process of being Americanized. If I were to remain in this country without him, I would surely lose all my Chinese in time. In that case, I would no longer be Chinese, because I had been taught and believed that it was essential to know the language in order to be Chinese. But I was also excited by the prospect of returning to China, a country which I barely knew. It was the proper thing for a patriot to do. The war did not concern me. As for leaving America, I felt I was going to come back to study here again in the future. I liked America, but like other countries I had lived in, Turkey, Austria and Germany, America was then, to me, a foreign country. It never occurred to me that I might become an American immigrant, as I later would. That possibility was not considered.

So my answer to his query was, "I would like to go back to China with you." In his book[2], Father notes that he was surprised but pleased with my response. He accepted it and he acted on it.

About going back to China, Father also had to consider Grace's desires. She had lived unhappily with us in China, Turkey and Austria for six years. She did not like living outside the United States, largely because she was unable to make friends, adapt or learn foreign languages. Father states in his book that while in Turkey she planned several times to leave for the United States, but she did not go through with her plans. She did not like living in China when there was peace, now that there was war, she definitely did not want to accompany Father back to China. She decided to remain in the United States. She wanted to move to Miami, Florida, where her good friends, Edna and Ruby Friedman, had moved from Brooklyn. One day, Father and I went with her to the Grand Central Station to see her off. I do not recall any tears at our leave taking, even though it was the first and last time I said good bye to her after seven years.

After we got home, Father told me that he and Grace had decided to separate, permanently. He found that they were incompatible. They were endlessly arguing and bickering. She had played the role of being my own "mother" since I was seven years old, but now he spoke with me for the first time about my own deceased mother, thereby denying the lie that he had upheld for eight years. I then felt free to tell him about her behavior while she and I were in the United States in 1938–1939. She was seeing a male friend, who often frequented our apartment. At the time, I was outraged by her disloyalty to Father, but because I respected her as "mother," I never told on her or criticized her. Father said he was aware of her behavior.

Up to that time, Father had been to me a loving but fairly detached parent, who was accustomed to obedience. He spent a lot of time teaching me, primarily Chinese, but we had few intimate conversations. The moments when we were close and informal together were when he was relaxed and told me one of his stories. The above two conversations, in the spring of 1941, were unprecedented. They had for me a defining importance. They were like my Bar Mitzvah, which separated my childhood from manhood. From then on, I felt that I was "grown up" because my father gave me the right to decide, or at least to participate in deciding, my own future. He also confided in me his personal problems. And we accepted the truth about my mother. I felt for the first time I was his friend, in addition to being his son.

Father and I had left the United States in August, 1941. We sailed for China from San Francisco to Hong Kong. Our intention was to fly together from Hong Kong to Chungking. However, in Hong Kong, Father met one of his schoolmates from Yale in China — Tsau So-Yen. At that time Mr. Tsau was a teacher of chemistry in one of Hong Kong's women's middle schools. He suggested that it would be better for my father to leave me in Hong Kong, which then was under the British and everyone thought it was safer. I could board at one of Hong Kong's Chinese Mandarin speaking schools, Ling-Ying Middle School. Its headmaster was a Chinese Christian, Hung Kao-huang, who had a graduate degree in education from Stanford. A number of Mr. Tsau's students were employed as teachers in the adjoining primary school section of the middle school and he promised that they could help take care of me. Mr. Tsau invited four of them to have lunch with Father and me, and we became acquainted. They were attractive single women, twenty five to thirty years of age. Two of them, Lee Chiau-Tzuan and Tai Kuo-Shin, would especially befriend me. Father was a man of quick decision, and after hearing about the possibility of leaving me in Hong Kong in safety, he decided that this was a good solution.

Father visited my classroom before he flew to Chungking. We were there alone. He went to my desk and pored through my textbooks. The textbooks in use were approved by the Ministry of Education in mainland China and published by the Commercial Press. They were used wherever there was Chinese education. These textbooks were familiar to us, as they were the same as the ones he had transported abroad when he taught me in Turkey, Austria and the United States. I showed him the lesson in Chinese that we were studying. It was a short essay in classical Chinese. He looked at it and gave me the book and started to recite the entire essay from beginning to end. I was dumbfounded, and also felt a bit sad, because I could not imagine that I would be able to catch up to him in that way. I have indeed never caught up with my father. In his day, the educated class in Hunan emphasized a great deal of classical Chinese. They practiced reciting the classics before they were teenagers. Aptitude for this type of training is very much like learning foreign languages. It is better done when one is young. As one gets older, recitation becomes more and more difficult, as shown by the example of the essay of that day. After he left, I tried to memorize the piece. After a few days, I did it with some effort. What I memorized one day was forgotten a week later. Even if I memorized it after a week, there is no way I would be able to remember it today, sixty years later. I have forgotten the piece completely; I don't even recall its name. I cannot recite the way my father did, reciting from memory at middle age what he learned as a boy. Of course, rote reciting is an old method and is no longer practiced. Even he did not enforce it when he taught me Chinese. Still, some things that could be achieved in the old days simply cannot be repeated nowadays.

When I started at Ling-Ying, I joined the third grade of junior middle school (equivalent to ninth grade). The challenges to me were many. The most difficult subject was Chinese. Although I had been abroad for six years and father constantly taught Chinese, I was still far behind my classmates in Hong Kong. Here I have to interject how difficult it is for a Chinese child to go to a Chinese school after so many years abroad. The Chinese language may be great, but it is extremely difficult. To be able to read and write those ideograms fluently requires years of intensive training. This is one reason why Chinese children who were educated abroad rarely attend Chinese schools when they return. They usually attend an English speaking school. There were many such schools in Hong Kong, but it never occurred to me to even think about transferring to one. After all, the main motive for my returning to China was to study Chinese. One of the things that I had to do was to write a composition on Chinese rice paper with a brush. The characters had to be written within the little squares printed on the paper. I was unable to squeeze the characters into those squares, because my ability to write was inadequate. I had to practice calligraphy assiduously with the brush, until I was finally able to fit my characters in the squares. Throughout my secondary school education in China, where I attended four different middle schools in Hong Kong, Guangdong, Hunan and Chungking, one of my prime concerns was catching up in Chinese. In Hong Kong and Guangdong, I had the added problem of different dialects. Father brought me up in the Hunanese dialect, which is close but not identical to Mandarin. Ling-ying was a Mandarin speaking school, but most of the students were Cantonese, and the *lingua franca* was Cantonese. It took me a few months to learn Cantonese. In the summer of 1942, I returned to Yiyang, and attended Sin-Yi middle school, the alma mater of my father in the old Norwegian Lutheran mission, now run by Chinese. Schools in Hunan and Guangdong are different in many ways. Hunan is further in the interior of China and is less influenced by the West. Even though both provinces used the same textbooks, certified by the Ministry of Education, there were differences in the teaching of some subjects. The standard of Chinese was especially high in Hunan, which was something I felt very keenly because of my low level. Before attending modern middle schools like Sin-Yi, many students had already studied classical Chinese at home in the countryside, so that they had some grounding in basic classical Chinese of the type that was not available from textbooks in our classrooms. In addition, they also practiced calligraphy privately at home so that they were accomplished in this art form and they could write beautifully. At that time people's health was so poor that there were many cases of serious illnesses, especially tuberculosis, among the classmates. Two classmates died while I was there. I learned about reverence for the dead in China. I had the experience of honoring my mother at her funeral. But this was about dead friends. We had memorial services in our classroom for each of the dead classmates. Part of the ritual was displaying memorial scrolls that were composed and written with a brush by classmates and hung on the walls of the classroom. These scrolls, usually couplets, were written in classical, stylized Chinese. I was taken by surprise, because I not only could not compose or write such scrolls, I even had trouble understanding some of them.

I had a further learning experience concerning Chinese at Nankai Middle School in Chungking in 1942, where I boarded for a year and half. Classical Chinese was not emphasized like at Sin-Yi. But a Chinese teacher, Tau Kwan, taught me something. Tau was a dashing, handsome, young man who wore a flowing, long Chinese gown. He said that the purpose of language was communication, to express thoughts or feelings, and not for adornment or showing off. Clear thinking and good writing go together. In composition, he emphasized the importance of clear and simple expression in the vernacular. He gave me a new perspective of appreciating literature, in Chinese or in any language. He emphasized the importance of reading and understanding novels. He brought me back to what Frau Burger had taught me about Goethe in Vienna. It was also at this time that I began to read serious novels in Chinese and English. For example, I read and enjoyed Louisa May Alcott, Dostoevsky's *Crime and Punishment*, and later, Charles Dickens in English. I thought of the lessons I had with Frau Burger about Goethe. By that time, I had begun to appreciate great writings in any of the three languages I knew. I was building a universal as opposed to a national point of view.

Tau Kwan's views on literature go back to the great Chinese literary renaissance of 1919, led by Hu Shih (1891–1962). This was part of the great modernization movement, called the "May Fourth Movement", championed by Chinese youth and led by professors of Beijing University. The date coincided with the date of the signing of the Treaty of Versailles. China was one of the victorious Allies, but because of secret treaties made by the Allies with Japan, she had to cede to Japan the Chinese city of Tsing-tao, which was being returned by Germany after her defeat in World War I. The promotion of "Democracy" and "Science" were other slogans of this movement. Hu was the primary leader of the literary aspect of the movement. He went through all the great works of Chinese literature and showed that the meaningful works were written in the vernacular, as opposed to the classical written language. Only the vernacular, spoken, common language could fully express real feelings and thinking of authors. This is consistent with the development of literature in all modern European languages, all of which were developed fairly recently, replacing ancient Latin or Greek, which Hu viewed like classical Chinese. Hu's position has been fully vindicated by subsequent events. Vernacular Chinese has become accepted as modern written Chinese. Expression in China is no longer constricted by classical Chinese as it was before 1919. Hu inaugurated the renaissance of modern Chinese literature. Still, written Chinese remains strongly influenced by the classical Chinese. The richest source of idioms is classical Chinese. It is impossible to appreciate certain aspects of Chinese literature, like poetry, in which the Chinese excelled, without reading or reciting classical Chinese works. I am a living example of the correctness of Hu's theories that vernacular Chinese should take precedence. I became competent in vernacular Chinese after five years of secondary education in China. But because I was not able to master classical Chinese during that time, excellence escaped me. I still feel that mastery of Chinese is best achieved with a solid foundation in classical Chinese, which my father had. I regret that I was unable to catch up to him in this respect, even though it was unnecessary to do so. Perhaps I am a bit jealous.

Besides Chinese, I had trouble in all my classes in Ling-Ying except English, but the more immediate problem was mathematics. We were taught plane geometry in the third grade (equivalent to ninth grade) in Hong Kong. This was a subject I had never been exposed to when I was in high school in Brooklyn. Mathematics is a subject that must be understood from its basics. One must start in the beginning and learn it in sequence. Fortunately, I was greatly helped by one of Mr. Tsau's students — Lee Chiau-Tzuan. She was a petite, quiet, prim and highly efficient young lady. I developed a crush on her. I fantasized that she would have made a marvelous wife for my father! She took time out every afternoon to review plane geometry with me so that I could catch up. In the beginning, she would spend a great deal of time explaining to me every theorem and its proof. After I got the idea, I was able to quickly go through the book, essentially by reading through it with her occasional help. After a few weeks of work together, I was up to the level of my classmates. However, we still kept our appointments and met together. We just changed the subject that she would teach me. She taught me Cantonese and I reciprocated by teaching her English conversation. In this way, we became friends.

1941 was not only an eventful year in my personal life, it was also an eventful year for the entire world, and Hong Kong was the focus of one of these events. I had been boarding at Ling-Ying Middle School not more than three months when Japan suddenly attacked Hong Kong after bombing Pearl Harbor on December 8, 1941. After a surprisingly short battle of just a few days, during which we underwent bombardment by Japanese artillery, the defending British army collapsed and Hong Kong was occupied. We were suddenly completely cut off from the outside world. The school closed its doors. Those of us, who had no homes, teachers as well as the students, took refuge and huddled together in the dormitory of the school. We took care and depended on each other for our everyday living. Some female primary school teachers, including Tai Kuo-Shin, one of Mr. Tsao's students whom I knew, moved in with us. Since there were not enough beds for all, we slept next to each other on bed rolls on the floor. Tai and I were neighbors. She was a cheerful, rotund type of person who was a lot of fun. She was intent on my teaching her English conversation. At night, lying next to each other on the floor and before falling asleep, she wanted me to tell her stories in English. I told fairy tales from Grimm and Anderson. The situation persisted for about two months. Then under the leadership of two teachers, Zhou Yiau and Wu Yun-yiao, a group of us organized and planned to leave Hong Kong for occupied Guangzhou on our way to free (unoccupied) China. Zhou was a history teacher who taught our class . His wife and two boys were with him. Wu was a civics teacher; whose wife was with him. Zhou was the more forceful leader. He was a Cantonese who had gone to college in Beijing, and who spoke perfect Mandarin. We looked up to him.

About a month and half after the Japanese occupied Hong Kong, a city that seemed to have died with the occupation came to life again. The shops remained closed but myriads of peddlers emerged in the streets. One day in January, many of us strolled and shopped in an open market with lots of hustle and bustle. There were vendors of all types, especially of foods. I was attracted to buying some Chinese bean-cakes and ate them. Flies were hovering around, but I paid no attention. A few days later, I was stricken with a severe case of bloody diarrhea with abdominal cramps. Unlike the usual case of diarrhea, I was so debilitated that I could barely get up from bed. They isolated me on a bed in a single room. I had to go to the bathroom ten, fifteen times a day, but often there was only tenesmus, blood and mucus. I felt seriously ill. I probably had either shigellosis or amebiasis. There was no physician or medication. Our date of departure for Guangzhou was set in a week. Increasingly, I feared that I would not be able to leave with the group. What would I do if I were left alone in occupied Hong Kong? Fortunately, on the day of departure, my diarrhea relented, and I was able to hobble out of bed and leave with the group. The first stretch was fortunately not strenuous. It was by boat a few hours from Hong Kong to Guangzhou. In Guangzhou, we piled into the large house of a friend of Teacher Zhou. They were very kind to accommodate us. There I recuperated.

For about a week, Teacher Zhou planned for the second stretch, which was the most hazardous one as it involved leaving occupied China and entering free China. The only way was to walk along unknown cow paths in the fields. Since no one knew the way, we had to engage a guide. Teacher Zhou found a middle aged peasant woman, who seemed strong and able. She came to orient us about the trip, bringing with her a bamboo basket with a false bottom. She showed us the bottom consisting of two layers of woven bamboo, and she suggested that we put our money between the layers to hide it from potential robbers. We gave her our collected bank notes and she hid them. After she left, we fell to wondering about her scheme. It seemed that we were at her mercy since she knew where all our money was. Discretion being better part of valor, we opened the false bottom, and each one reclaimed his own notes. I hid mine in a cloth belt I wore around my waist. Others hid theirs in various and sundry ways.

From Guangzhou we proceeded to walk through no man's land into free China. First we had to go through several rounds of inspection by the Japanese army. As there was a crowd of about a hundred people waiting for inspection, we were delayed. When we started to walk through no man's land, it was already dusk. In the beginning there were about twenty or thirty people walking together, but gradually with the onset of darkness our numbers dwindled to just our group of about ten people. Soon it was pitch dark and we had not yet arrived. Our guide said we better stop at an abandoned roofless temple we hit upon to spend the night. I am sure she had not anticipated this. Just as we were lying down on our bed rolls to go to sleep, four or five men descended upon us brandishing guns, revolvers and flashlights. They said, "We are guerrillas, don't be frightened." Then they asked each one of us to stand up as they searched us and took whatever money or valuables they could find on us. I was lying near a wall and farthest away from them. As I was waiting for my turn to be searched, the thought occurred to me that I could just step outside the open window, which was remarkably lowly situated, and flee. But I stayed put. Why? Was I just naturally timid and circumspect, or did I have a streak of audacity and risk taking? Finally, a man came to me, and my opportunity to flee vanished. He immediately found, conveniently attached to my belt, all my money. He just asked for the whole belt. He then asked me to take off my heavy overcoat I was wearing. He took it.

I said, "May I have my glasses which are in the inside coat pocket?" He said, "No!" Within minutes, they were gone.

I was suddenly impoverished, half blind and cold. No one of my family was near by for comfort.

But this experience also taught me something about Chinese culture, which can't be learned in the classroom. The relationship among Chinese students and teachers has a closeness that is not paralleled in the West. They remain close friends even outside of the classroom, and beyond the time when they are in school together. They take care of each other during times of need. While I did not have any more money, I felt instinctively I would be taken care of by the group. I was like in a big family. I did not ask them, and they did not have to offer.

My feeling was quite different just a few months before.

In September of that year Father and I said goodbye. We were standing in the playground of Ling-Ying Middle School in Hong Kong. Ling-Ying was located on a promontory on the shores of Causeway Bay, on the eastern side of Hong Kong. We were standing and talking on the top of a flight of stone stairs, at the edge of the playground that led down to the street. He was flying to Chungking the next day. I would be left alone in this Chinese boarding school for the first time, essentially in cultural shock coming from a school in America. He told me to take care of myself, and to write. Despite my fulfilled wish to return to China, at that moment I felt miserable, lonely, abandoned and neglected. He and I embraced so tightly that our bodies were tensed. Our cheeks touched each other. Tears swelled up in my eyes, as they did in his. We said "goodbye" to each other. I watched him going down the stone steps. I followed him and waved to him until I could no longer see him. Then I went to a private corner in the playground and wept bitterly.

On the day that father and I parted, I felt that I had been thrown by myself into a strange environment. I felt lonely and sad. In the old days, when I was abroad and in a foreign school, I was always regarded curiously because I was the only Chinese in the classroom. Now that I was in a Chinese school, and I was Chinese like the rest, I was still considered strange because while I looked Chinese, my behavior and way of talking were different. Whether in China or abroad, children's prejudices are remorseless. It is a characteristic of children, particularly as a group, to love to tease. I acquired a nickname in Cantonese, "so-tzai," which meant the "awkward one."

Now, after the robbery, I was no longer "so-tzai." I felt I belonged. In fact, I later acquired another nickname, "second brother". That is because, among the group, I was considered second in seniority among our group of students. Since families are the most important unit in Chinese society, according to Confucian ethics, other social associations assume appearances of a family to signify closeness.

After getting into free China, our group of refuge students and teachers took a house-boat on the West River and headed north for Ku Kong. (The West River is one of the three major tributaries of the Pearl River, the major river of Guangdong, which empties into the China Sea near Guangzhou.) This trip took about a week because our house-boat had to be pulled up stream. Despite my sadness, life on the boat was pastoral and tranquil. The scenery was idyllic along both shores of the West River. They were heavily wooded by dense bamboo groves. The water was crystal clear. This is a scene that remains in my mind that is quintessentially Chinese. It was like a Chinese painting. After we arrived in Ku Kong, we were sent to the refugee center for Hong Kong students where we were very well treated. All refuge students immediately received a padded cotton coat, before we were treated to a hot meal. The weather is quite cold in February in northern Guangdong, and after I had been robbed, I was devoid of a overcoat and was cold most of the time. When I put on the padded coat, I instantly felt a glowing warmth permeating my body that is impossible to describe. It is an experience that most people will never undergo, in this day of abundant creature comforts.

In Ku Kong, I telegraphed my father in Chungking. He wired back and he told me that he was getting married in a few weeks. He sent me a fairly large sum of money, two thousand Chinese dollars. This was very welcome, since I was without any funds and had been depending on my friends. I was happy to share with them part of what I received. I was particularly helpful to an overseas student, who was also robbed clean. We shared and shared alike. At that time, there had been many Christian Chinese middle schools in Hong Kong, and many of their students and teachers went to free China. A temporary compounded middle school, called Ge-Lien (Christian associated) middle school, was established in an American Presbyterian compound in Lien-Hsien in northern Guangdong to accommodate the refugee students. Lien-Hsien is a remarkably scenic spot, with a crystal clear stream and verdant woods interspersed with rice paddies. I transferred to the third grade (ninth grade) of the junior middle school. The third grade is the graduating class. In June of that year, I graduated from the junior middle school at Ge-Lien. Originally I thought I would then reunite with my father in Chungking, but in April 1942, he was again transferred, this time from Chungking to Washington D.C. So he was no longer in China and I was left again to fend for myself. I decided to leave northern Guangdong for northern Hunan and to go back to my hometown, Tau-Hwa Lun. I attended grade one of Sin-Yi senior middle school (equivalent to tenth grade). This was the school that my father had attended, where the principal and many of the teachers were his schoolmates.

Graduation class from junior high school at Christian Associated High School in Lien-Hsien, Guangdong, 1942. Monto Ho is tenth from left, second row. Notice absence of glasses! Teachers are seated at the front; the three Americans are from the left, Mr. Robert Mongomery, Mrs. and Mr. Fuson.

I stayed with the family of my paternal aunt, Shau-hua, with whom I was reunited after an absence of seven years. She and her two daughters and son shared a house with the family of a faculty member of Sin-Yi. Her husband, Wong Tao, was a civil servant working with the Department of Education of the Hunan provincial government, which had evacuated to Leiyang, in southern Hunan, from Changsha because of the war. They had spent their entire lives in Hunan. One daughter, then twelve years old, went to Sin-Yi middle school. The other daughter, who was seven, was left at home. The son was an infant. Like her own mother, Shau-hua had eight or nine babies, of whom only a few survived. Shau-hua had only graduated from middle school. In appearance and temperament, she resembled Father. She has his broad face and fair complexion. Like him, she was an excellent speaker and racounteur. Both of them dote on people they love. She welcomed me with open arms, and catered to me and pampered me during the whole time I was in Yiyang. Her devotion to Father was manifest. And it was mutual. For example, throughout all those war years, when she had to move repeatedly with the barest of essentials, she carried with her everywhere she went Father's heavy German books from his graduate studies in Munich. As for him, he gave money to her throughout their lives. During the Chinese New Year holidays, she took me to the countryside to visit the old family homestead of the Ho family. Here were some cousins of hers and father's. Visiting relatives during this holiday is a Chinese custom. For the first time, I lived among Chinese country folk in the villages. I observed their customs. I slept in their houses and ate their food. This was the real China! I felt an affinity with aunt Shau-hua because of her closeness to Father. This was the first time I felt at home after returning to China.

Still, I could not condone the way she treated her younger daughter. For some reason, she hated her. I didn't even know her name, she was simply called "little girl". She was treated like a slave. She did not eat with us. She would get a bite to eat off and on while standing in the kitchen. All she did were menial chores. She was not allowed to study. Nobody talked to her decently. She was a small girl with a limp. Apparently her left leg and thigh had been burned, and there was much scar tissue. I often wondered how she got the burn. Every few days she would be beaten by her mother. They were brutal, merciless beatings. The pity of it was that she did not even dare to cry. She merely let out stifled whimpers.

Here was a case of flagrant child abuse, but it was made worse because it was in China. The Confucian ethic extols the family. The relationship between parent and child is supreme. Part of this ethic commands the obedience of the child to the parent; so much so, that the rights of the individual may be submerged. I have often felt that in over emphasizing filial piety, the Chinese are over emphasizing a one way relationship; that of child to parents. It is not uncommon for parents in China to expect and even demand support and care from their children and daughters-in-law. The idea of rights of children would be foreign to Confucius, although there is the expectation that children be cared for by parents. The wonder of aunt Shau-hua's case was that her cruelty seemed to be above rebuke. No one dared intervene for "little girl"; not only that, no one even dared to criticize. The power of the parent over the child seemed absolute. No one stood for "little girl's" rights as a human being. Ordinarily, no mother would be that cruel to her child. It is not natural. But unnatural things happen in the best of worlds. When that happens, there must be other commandments or rules that hold them in check. The rule that I held to, was the commandment to love your neighbor. If you cannot love your daughter, you should at least love her as your neighbor. China is a country where one's neighbor is a stranger. Aunt Shau-hau could not love her neighbor, even though she was baptized as a Christian. I was puzzled and disappointed in her.

Cruelty is not a stranger in China. This was but an instance within our family. I am afraid it was also a reflection of Chinese society. I saw too many violations of human rights, along with poverty, ignorance and dirt in China. Is it related to the culture I so admire? Probably yes, unfortunately.

One day in the fall of 1943, Sin-Yi middle school was visited by a group of Western correspondents. They were on their way to Changde, where the Chinese claimed a victory over the Japanese. The leader of the group was Wei Ging-mon of the Information Service of the National Military Commission in Chungking. He addressed us students at our regular Monday morning assembly. As I listented to him, I recalled that my father had just returned to Chungking from the United States. We talked about my going there to join him. The problem was, how would I get from Yiyang to Chungking? Father had no immediate solution. Communications within the vastness of interior China was difficult to begin with, but it was well nigh impossible during the war. Main highways and railroads were either destroyed by the Japanese or scuttled by the Chinese. There were air flights from Guilin to Chungking, but they were not available to ordinary people. One had to have connections. In any case, it was not easy to get from Yiyang to Guilin, which was 400 miles away in another province, Guangsi.

It occurred to me that since the group came from Chungking, they must have means to return to Chungking. After the assembly, I introduced myself to Wei as the son of Ho Feng-Shan. Perhaps he knew him and would he help me get to Chungking? Mr. Wei did not tell me whether or not he knew my father, but I surmised that at least he knew of him. He said that they would have to go to Changde first, but within ten days or so, on their way back, he was willing to take me with them. But he cautioned me that he could only take me as far as Guilin. He had no way to provide for me a seat on the flight from Guilin to Chungking. I would have to fend for myself from there on.

This brief conversation at least solved half of my transport problem. I told Aunt Shau-hua, and she prepared for my departure with sadness. For in those days, once separated, when would we see each other again? Indeed, it was almost forty years later that I saw her again during a visit in Chengdu, shortly before her death.

I might remark that two years after returning to China from the United States, I had been transformed from an American boy into one of thousands of poor students in interior China. Students underwent paramilitary training in high schools. Our hair had been cropped bald, as common soldiers. We wore uniforms made of cheap cotton cloth the upper part of which buttoned down the entire vertical length (so called Lenin, Sun Yet-Sen or Mao Zedong costumes). We wore soft cloth shoes. Leather shoes were considered a luxury. I did not wear them even if I owned a pair. My appearance and position was such that I did not feel any affinity with the American correspondents. I did not feel free to "fraternize" with the Western correspondents. Mr. Wei did not introduce me. We seemed to belong to different social classes.

Still, I was glad to be able to join the group for the trip. The first part was from Yiyang to Changsha, the provincial capital, a distance of 60 kilometers. Since the highway was destroyed, it took two days of walking. About twenty sedan chairs were provided for the group, a great luxury. I was assigned one with two porters to carry it. The porters were able to make good time, as their way was to trot rather than walk. Trotting improved walking efficiency. In the beginning, it was fun to sit in the sedan chair and enjoy the surrounding scenery. But as time went on, I observed the sweat streaming down the faces of the porters, especially when they were carrying Westerners who weighed over two hundred pounds. Empathy got hold of me, and I felt disgraced that a healthy young man like me should be carried. I told my carriers that I wanted to get down and walk, and they could have a rest. This went on for a few hours and we then had a lunch stop. After lunch, I looked for my sedan chair and porters. They were nowhere to be found. They had vanished! It turns out that these sedan chairs and the porters were conscripted by military force, as Mr. Wei's organization was the all powerful National Military Commission. And desertion from conscription was a well known phenomenon in China! I was told that no replacement of my sedan chair was possible. There was no alternative but for me to walk, or trot. By the end of the first day, my feet were already sore and swollen. By the end of the second day when we arrived in Changsha, there were open blisters on my feet.

We were put up in an exclusive hostel in Changsha. After dinner, as was my wont, I went out of the compound for a walk. I came back after an hour, but I was stopped at the gates by the guards. How could a shabbily-dressed, poor student like me belong to this group with Westerners? I felt humiliated. Only after confirming my identity with the leadership of the group, was I allowed in. One does not realize in these days of reverse snobbery that one's appearance and clothing could relegate one to a stigmatized "lower class" in the old days.

I felt deeply, not for the first time after my return to China, for the oppressed and the downtrodden. Everywhere we went, we saw suffering. I saw it in the laborers groaning under their heavy burdens and farmers straining at their plows. These were the luckier ones. How many were suffering under injustice and oppression, like Aunt Shau-hua's "little girl"? Thousands were forcibly recruited in the army. Millions were poverty stricken, not even strong enough to beg for their meals. The depth of my feeling for China's misery was a factor in my later decisions in college for my career.

In Changsha, we took a train overnight all the way to Tu Shan, near Guilin. There I left Mr. Wei and his group who flew to Chungking, and I was left to my own devices. There was a highway built through the 400 miles of mountainous terrain from Guilin to Chungking, but there was no bus service. The only way to get a ride was to be connected with an agency or company that serviced the many trucks on the road. Alternatively one could be a "yellow fish", or stowaway in a truck, if the driver would let you. In any case, all trucks must stop at an inspection station going out of town. That is where I headed the morning after I got to Guilin, carrying my wooden black suitcase with all my earthly belongings. Many potential passengers were waiting at the station. I got to talking with a middle aged man who seemed very confident. He found out who I was and where I wanted to go. There were two types of trucks plying the road. The usual gasoline engine empowered trucks were in the minority. Most of the trucks had a type of engine that I only saw in war torn China. There was an auxiliary stove mounted on the side with charcoal as its fuel, and a two stroke engine that consumed carbon monoxide. This contraption saved gasoline, of which there was a great shortage. The only problem was they could not go faster than 30 or 40 miles per hour. And they had a particularly tough time climbing steep grades. My friend let several charcoal burning trucks go by. When an empty regular gasoline burning truck came by, he went to the driver, showed him some documents, and then he signaled me to follow him to the back of the truck, which was half empty. We had the whole truck entirely to ourselves! It was a two day trip even as we wheezed through the mountains. Soon we were near Chungking. He finally told me that he was a secret agent working for the National Military Commission. He had authority to board any truck. I was duly impressed, but somehow I developed a fear of him. When the truck stopped at an inspection station across the river from the city, and he had gone to the bathroom, I hoisted my suit case off the truck, and hustled to the ferry to cross the river. I hired a tricycle carriage and found my way to my father's house. Were they surprised when I arrived! Later I felt badly I had not thanked my friend. Very likely my fears were ungrounded, and he was just doing me a good deed. My trip from Yiyang to Chungking took a week. After that, I felt completely confident traveling in war-torn China.

Having finally arrived in Chungking, with the joyful reunion with my father and new stepmother, I was later admitted, after passing an entrance examination, to Nankai Middle School. Nankai Middle School was a school that originated in Tianjing and was founded by Chang Bo-Ling. It became a very well known and respected national middle school. There were many differences between Nankai and the schools I attended in Hong Kong and Hunan. Nankai is a school that had achieved an outstanding national reputation, which it maintains until this day. In addition to an excellent academic tradition particularly in mathematics and the physical sciences, Nankai also emphasized extracurricular activities such as sports and the arts. A previous premier of communist China — Zhou En-Lai, was a graduate of Nankai and he was well known for playing the role of women in plays that he acted in. I studied for a total of one and half years at Nankai from the second semester of the second year to the end of the third year when we graduated. Another characteristic of this school is that classmates were particularly close. Our classmates have remained in touch with each other to this day. We have a class alumni bulletin which is published several times a year. In contrast, the alumni of my Hong Kong and Hunan schools do not maintain such close contact.

chapter
| 4 |

REMEMBERING FATHER: FENG-SHAN HO

O n January 23, 2001, my sister Manli and I went to Jerusalem to receive the "Righteous among the Nations" award for our father from Yad Vashem. The award is the highest honor that Israeli government can bestow on a non-Jew. Yad Vashem is the "Holocaust Martyrs' and Heroes' Remembrance Authority" of Israel.

Father, Grace and I arrived in Vienna, Austria from Ankara, Turkey in the summer of 1937. Father was assigned to be the first secretary and charge d'affairs (chief of the mission) of the Chinese Legation in Austria. This was a promotion from his job in Turkey. He became avidly involved in the social and cultural life of Vienna. Austria had 185,000 Jews of whom 166,000 were in Vienna. Most of them were merchants, professionals, intellectuals, and artists, and many were active in society. Many became our friends.

Hitler annexed Austria overnight on March 11, 1938. Austria being no longer independent, the Chinese Legation was dissolved and was transformed to the Chinese Consulate-General. Father was appointed Consul-General. We were in greater Germany now. The Germans renamed Austria as the province of "Ostmark". The Germans enforced their anti-Jewish, discriminatory laws, which made it difficult for Jews to earn a living. Many Jews were rounded up and sent to concentration camps, usually Dachau. Although few were killed at the time, most Jews felt the pressure and were eager to seek refuge outside of Austria and Germany. At that time, a conference of thirty-eight countries of the League of Nations discussed ways to help the Jews. This "Evian" conference reached no conclusion as to how the Jews might be helped. In Vienna, most Western consulates were no longer issuing visas to the Jews. The British and the French consulates stopped issuing visas. The Swiss consulate would issue a visa, but it had to be marked with the letter "J," signifying that the recipient was a Jew. The quota of visas for Jews in the American consulate was very small. Under these circumstances, the Chinese consulate, under Father's jurisdiction, began issuing visas to Jews so that they could leave Austria for Shanghai. This fact became well known. It issued visas from 1938 till 1939. Father continued to do this despite orders in late 1938 from his superior in Germany, the Chinese Ambassador Chen Chieh, to desist. Chen was afraid that what Father was doing would harm China's warm relations with Germany.

Whoever had a visa for a foreign country was allowed to leave Germany, even though he may not have gone to the country designated. It was not known at the time that whoever received a visa was saved from annihilation, because almost every Jew who remained in Nazi Germany would subsequently succumb. After 1943, they were all systematically killed.

There are many stories of Jews who were helped by Father's visas.

The following is in his autobiography. Shortly before the so-called "Kristallnacht", November 9, 1938, a young Polish Jew in Paris assassinated the third secretary of the German embassy. For this, Hitler ordered the rounding up and interrogation of Jews all over Germany, which included Vienna, Austria. The Nazi storm troopers went on a rampage and wantonly destroyed Jewish shops and dwellings. The large amount of broken glass that resulted was the origin of the term "Crystal Night".

On that particular day, Father had an appointment to see his Jewish friends off. Mr. and Mrs. Rosenberg were to leave Austria. During the day, Mrs. Rosenberg had called Father to tell him that Mr. Rosenberg had been arrested. She suggested that in view of the present situation, it would be better if he did not come. But he insisted on keeping the appointment. Soon after arriving at their house, two plainclothesmen arrived at their house and demanded entrance.

The maid who let him in said in German, "Grüss Gott," which means, "God greet you," or "good day," in vernacular German.

The cold plainclothesman said, "Say 'Heil Hitler,' and not 'Grüss Gott.'" He meant that one should say "heil Hitler" instead of "good day."

Mrs. Rosenberg said that we have already been searched today. The plainclothesman then said, "Well, now you'll be searched again."

At this point, Father was sitting and smoking a cigarette and wondering what to say if he were questioned.

Sure enough, one of the plainclothesmen asked, "Who are you?" and Father answered, "If you will first tell me your identity, I'll be very happy to tell you mine."

This rather cold response surprised the man, and he turned to Mrs. Rosenberg and asked, "Who is this man?"

Mrs. Rosenberg answered, "He is the Chinese Consul General."

In those days, diplomats had a great deal of prestige. Therefore, the policemen quickly left the house without searching it. Soon afterwards, Mr. Rosenberg was released and the couple eventually escaped to Shanghai with the visas that the Chinese Consulate-General had given them.

Lilith-Sylvia Doron, now living in Israel, met Father accidentally as both watched Hitler entering Vienna, on March 11, 1938 — a time when the Nazis physically assaulted the city's Jews.

"Ho, who knew my family, accompanied me home," says Doron. "He claimed that, thanks to his diplomatic status, the Nazis would not harm us as long as he remained in our home. Ho continued to visit our home to protect us from the Nazis."

When Doron's brother, Karl, was arrested and taken to Dachau, he was released thanks to a visa issued by the Chinese consulate. Doron and her brother left Vienna in 1939 for Palestine[3].

Eric Goldstaub, now living in Canada, relates how in July 1938, he received Chinese visas for his entire family after spending "days, weeks and months visiting one foreign consulate or embassy after the other trying to obtain visas for himself, his parents and relatives, numbering some 20 people".

The rush for visas assumed panic proportions after Kristallnacht, when thousands of Jews were thrown into concentration camps, only gaining release if their relatives produced visas or tickets for travel to other destinations. Gerda Gottfried Kraus, based in Canada, relates that after Kristallnacht, her husband waited in a long line for admittance into the Chinese consulate. Seeing a car approaching the consulate's gate, he thrust his application form through its window.

"Apparently, the consul-general received it, because my husband then got a call and received the visas"[3].

How did Father's persona come about? In going back, his entire childhood was spent in Tau-Hwa Lun, at the Norwegian Lutheran Mission. He was born in 1901, and lost his father when he was seven years old.

Grandfather was a poor intellectual. He was respected for his learning and his calligraphy. That is, he had a classical Chinese education of the Ching Dynasty, but he did not qualify in any of the state examinations which might have given him an official position and a comfortable livelihood. His early death made life difficult for his wife, my grandmother. She gave birth to eight children, only three of whom survived to adulthood. Infant mortality was very high in those days. The oldest was a daughter, who was married to an officer in the army. She was relatively well off. The two youngest children, who were more than twenty years younger than her, were my father and his younger sister, Shau-hua. They were brought up by their widowed mother, alone. Grandmother was a devout Christian with little income. She was a practical nurse in the orphanage of the Mission, a position which barely sustained her and her family.

Before going to middle school, Father was quite a mischievous boy during his primary school days, giving his mother all kinds of trouble, even though he was extremely devoted to her. He played with the neighborhood urchins and they often dared each other to do outlandish things. Most of them had become good swimmers cavorting in the Tse River. Unguarded swimming in China is not to be scoffed at. An excellent Norwegian swimmer and athlete, who was a missionary in Tau-Hwa Lun, drowned while swimming in the river. The dare on this occasion was to dive down one side of a sampan and come out the other side. These sampans were large heavily laden freighters which plied the rivers carrying cargos of various types. The hull of the ship would be about eight feet deep, but its width was what was misleading. Once underneath, one had to be sure of one's way and swim correctly and swiftly across its width in order to come up safely on the other side. Father dived down one side but got really scared while looking for its width to cross. He fortunately found it and safely reached the other side. He won his dare. But by the time he returned to shore, his concerned mother had dispatched someone to look for him. He went with him soberly home.

Although his family was poor, with the help of the Mission, Father grew up in its beautiful environment and flourished physically and intellectually. His education from primary school all the way through high school was received at this mission. The Chinese emphasize learning and intellectual achievement. Many people said of Father's scholarship, using a classical Chinese expression, "once read, always remembered". It is difficult to know what the truth was. There were many bright and intelligent people in olden China. But during the second year of middle school (at that time middle school consisted of four one year grades), he participated in a school-wide contest in English and placed first, despite being an underclassman. This prize made him well-known throughout the school, and his teachers, including the Norwegian headmaster, took note of him. He was able to enter Yale in China College in Changsha, the provincial capital, because he got financial help from the Mission.

Years later, I met one of Father's American teachers in Boston, when I was a student at Harvard College. Mrs. Theodora Ward was a New Englander, who had gone to Tau-Hwa Lun as a young teacher of English. I met her when she was in her sixties, and she was working on a book about the poet, Emily Dickinson. She frequently had me at her house for dinner and reminisced about Father. She said that, in the beginning, she had to correct the Norwegian inflections in his English pronunciation. This was common among students at Sin-Yi Middle School, as they were taught English by Norwegian missionary teachers. He was eager to go beyond his assigned work in English. He would study and learn obscure phrases in a book of phraseology that he borrowed from the library and tested her. So she had to be on her guard! He was not an ordinary student.

Yale in China College was founded by Americans from Yale College and certified in Connecticut. It was not a missionary school. This college existed till 1927, the year I was born. By that time, Father had graduated. Nowadays, after the Communists came to power in China, Yale in China was reestablished as the Yale China Association at Yale University. It no longer runs any educational institutions in China, but it assists schools it originally founded in China. It sends bachelor teachers of English to Yali (Yale in China) Middle School as well as Hsiangya (Yale in Hunan) Medical School; both schools in Changsha were founded originally by Yale in China. The five years of college in my father's life were happy times for him. He was not only an outstanding student academically, but he was extremely active in extracurricular activities. He was chairman of the college YMCA, the student association and perennially he was captain of the soccer team.

There was a brief period in 1927 when the Communists were in control in Changsha. Father had joined the Kuomintang, and was active on the college campus. When the Communists came into power, one of their prime targets was members of the Kuomintang, and people who worked with the "imperialists", which included the Americans at the Yale in China campus. Accused people were brought to trial at a mass meeting. The accusations were read in front before the accused. At the end of the trial, the punishment was meted out. It was finalized by people raising their hands. Being unequipped with microphones, people in the back had no idea what was being said and decided on in front. They just raised their hands when people in front raised theirs. Many were executed after such farcical trials. Under these circumstances, Father fled to Shanghai for six months, until it was safe to return to Changsha again. These early actions of the Communists made a lasting impression on Father. After World War II, when the Communists agitated for "democracy" and "freedom" while fighting the Nationalists in the civil war, Father never believed any of their propaganda. He always went back to his earlier personal experience. This may be one reason why he was so steadfast in his anti-Communism.

After graduation from Yale in China, Father worked in the provincial government of Hunan. He was recognized and patronized by the Governor, Ho Chien (1887–1956). In 1930, the province of Hunan provided two scholarships donated by a German trading company for study in Germany. Governor Ho wanted to award one of these to Father, but he insisted on taking the examination for the scholarships like anyone else. The result of the examination was that he was one of two examinees who got identical scores, and placed first in the entire province. Both were able to go to Germany for graduate studies. Father studied at the University of Munich for three years. He studied political economy and obtained his Ph.D. in 1932. In 1935, he joined the Chinese foreign service, to which he devoted forty years of his career. He retired in San Francisco, and died in 1997.

Despite his early favorable environment, it was difficult for a young man in remote Hunan who was poor and without means to go beyond his environment and achieve. The Chinese civilization has always had this particular characteristic, and that is that irrespective of the times or dynasty, whether it was war time or peace time, there seemed always a way for a person of merit to extend beyond his immediate environment and achieve. Even though the imperial examinations were abrogated with the end of the Ching Dynasty in 1911, Father's early years were punctuated by success in important examinations very much like success in the old imperial examination system. This is in the tradition of Chinese culture, that is "those who achieve excellence in learning become officials". It is one of the reasons for the longevity of Chinese civilization.

In 1935 we were on our way to Turkey, where Father had a new job with the Chinese legation. Grace, my father and I were traveling from Shanghai to Istanbul on the Italian liner named "Conte Verdi" of the Lloyd Triestino Company. The trip took three long weeks, with pleasurable stopovers in Hong Kong, Saigon, Singapore, Colombo (now Sri Lanka), Madras, Bombay, Aden and Port Said. There were many other children, seven-and eight-year-olds, on the ship besides myself. There were innumerable places on this gigantic ship of 20,000 tons where we could roam and play. I was having a glorious time. One day, my father discovered that my behavior was distinctly abnormal. Instead of my playful self, I was intentionally avoiding a Caucasian boy bigger than myself. No matter when and where, whenever he was around, I would slip away and avoid him. Finally he asked me why was I avoiding him. I told him that the boy was a bully and he picked on me and hit me whenever he could. Father told me that there are times in life when one has to stand up and fight for one's rights. He then proceeded to teach me a wrestling trick. He showed me how to stand facing the child but putting my left foot behind his legs and to push his chest backwards with my left palm. In this manner, the boy would be tripped and pushed to the ground. He spent about three or four mornings practicing this trick with me so that I would be facile in carrying it out. Finally, he said, "You are ready." Sure enough, one day the boy appeared again, intent on picking on me. This time I stood up to him and performed the trick that Father had taught me. To my pleasant surprise, I tripped the boy and he lay on the floor flat on his back. From then on, the boy never bothered me anymore.

The Chinese legation in Turkey had a large staff of about thirty and Father was the second secretary. Actually, China and Turkey were countries far apart with relatively little in common. There was really not much for the staff of around 30 people to do. There were two people among the staff who were exceptions. They were Father and his good friend Wang Pun-sun. Mr. Wang was the counselor of the legation, and he too, like father, was from Hunan. He was a soldier scholar who was very learned in both Chinese and Japanese. He was reputed to be among the one or two foremost experts on Japan during the later war with Japan (1937–1945). Both he and Father were scholars at heart. Together, these two people undertook a program of systematic writing. They studied and wrote about the new Turkey, which was then being built up under its founder Mustafa Kemal Atatürk (1881–1938), after her devastating defeat in World War I. They specialized on writing about Turkey's internal conditions and diplomacy. Because Father was versed in both English and German, besides Chinese, he was able to read a much wider scope of relevant literature than Mr. Wang. Much of their conversation was taken up by Father's discussing with Mr. Wang what he had learned in his readings. During the two years, the two men produced many articles for various newspapers and magazines. Father completed two books reporting on the social conditions, the economy and agriculture of the new Turkey. The names of these books were *The Politics,*

Economy, and Society of the New Turkey and *The Development of the Agricultural Economy of Turkey*[2]. The publication of these two books was negotiated with the Commercial Press, which was then the best known publishing house in China. However, the Ministry of Foreign Affairs in Nanjing intervened and prevented their timely publication. They felt that this valuable material should not be made public commercially but should be published by the Ministry of Foreign Affairs. The delay in its publication resulted in its total loss after July 7, 1937, which marked the beginning of the Sino-Japanese war (1937–1945). When Nanjing was occupied by the Japanese in 1938 and raped by the Japanese army with horrendous loss of civilian lives and property, this material was irretrievably lost; one of the many losses suffered by Father during the war.

We left Turkey after three years. Father was transferred to the Chinese legation in Vienna as first secretary. In 1937, the Sino-Japanese war had broken out and the war was very much in the news. Father had studied in Germany, and he was fluent in both spoken and written German. He considered it his duty to explain to the Austrian public the justification of China's actions in protecting itself from Japanese aggression. Since the latter part of the 19th century, Japan had pursued a series of aggressive acts, economic, political and military in nature, against China. The situation was similar to my being picked on by the bully boy onboard the ship. Father collected his speeches in German and published a book called *China Verteidigt Sich*[2]. Translated this means, "China Protects Herself." Besides public speaking and publishing, Father also engaged in debates in the press with the Japanese diplomats in Vienna about the pros and cons of the war. To an impartial mind, there could be no doubt as to the blame.

Father had a strong sense of right and wrong and fought for what was right. He was a consistent hard worker devoted to excellence. He was lively and loved fun. He was also a good looking man with an athletic build.

"Keep your back straight and your head high", he said to me repeatedly when I was a child.

A certain amount of "gravitas" was in his manner, and his presence in a room was easily felt. He had an easy smile and laugh. His loud laugh was hearty, good natured and contagious. He was an excellent speaker, who reasoned logically and eloquently. He had a temper, which occasionally erupted, but his anger was temporary. He got over it quickly.

His life was simple. He had no hobbies, and did not care to collect material goods. Although he was an economist, he was a poor manager of his personal finances. He smoked moderately until the age of 60. One day, after developing a severe cold, he gave up smoking temporarily, but he never resumed the habit. Drinking and gambling did not interest him. What he enjoyed most was conversation and friendships. He liked nothing better than to converse in depth with friends and relatives. They talked about mutual acquaintances, Chinese politics, international politics, history and daily lives.

All his life, he paid attention to physical exercise. When he was young, he excelled in athletics. While a student in high school and in college, he was perennially captain of the soccer team. Even when he became old, he did not forget to exercise. When 80 or 90 years old, he would take a brisk walk every day in the neighborhood of his house in San Francisco. I often accompanied him during these walks. They were times when we engaged in heart to heart conversations about our families, reminiscences, and politics. He would tell or retell one of his stories. They were about his past, people he knew and Chinese history. Father was healthy throughout most of his adult life. Except for minor colds, he was never seriously ill. He was admitted to a hospital for the first time when he was 92 years old because of a stroke on his left side. Even though he recovered to the point of being able to walk with assistance, he never fully recovered his mobility, and this illness eventually led to his death at the age of 96.

Father's diplomatic career extended from 1935 to 1972. His last post was the ambassadorship to Colombia. Thus, he served the Chinese diplomatic service for almost forty years. Some people may have the misconception that diplomats are fast-talking, socialite types of politicians whose main concern is outward appearances. Father was a sincere and dedicated person, who loved his country and was devoted to doing his work well. He also had the social qualifications of a successful diplomat. Whenever he arrived at a new post, he would make it a point to meet the important people in government and in society. He was quite unlike some Chinese diplomats who isolated themselves behind closed doors. His social activities were very much helped by his wife, my second stepmother, Huang Shao-Yun, who was an attractive and popular hostess.

From 1943 to 1947, Father was director of the Department of Information of the Ministry of Foreign Affairs. This was his first position among the leadership of the foreign service. He demonstrated his abilities as an administrator and leader of men. He got rid of the corrupt practice of members of the staff signing in without actually appearing for work. Like the veteran teacher that he was, he ran the department like a school. He paid attention to the advancement of his staff, but he would only reward those who were truly deserving. He worked his people hard. Thus, he wrote in his book[2],

> "These young people are truly deserving as they are very receptive to being led. As long as you treat them honestly, they will adhere to your leadership. Pretty soon, this department of seventy people was functioning as if it were of one mind. I have become like a university professor or like a commander of a small army." (p. 166)

Many years afterwards, when I met former members of his staff from that time, they still reminisced about those good old days when Father was an extremely popular Head of Department of Information. It collected information from overt sources from both inside and outside of the foreign service. Father produced the only English language newspaper at that time in Chungking. He also edited a periodical called "Political Life", consisting of articles of human interest. When I was at Nankai Middle School in Chungking, I wrote several articles for it. They were called "My Childhood Days in Vienna", and "My Primary School Days in the United States".

To me and my sister Manli, Father's love of conversation was expressed in his storytelling.

One of the most favorite periods for his storytelling was the period of the "Three Kingdoms" (190–317 A.D.). This was a period after the fall of the great Han Empire (206 B.C.–221 A.D.), when China was splintered into three states, each assuming it was entitled to the throne. The battles, diplomacy and statecraft practiced by the three adversaries, and the individuals involved, constitute some of the best romances and stories of Chinese history. One general, and later prime minister of the Zhu Kingdom (in present day Sichuan) was Zhu-Kuo Liang, who was famed for his battle tactics and strategy. He was a genius in the practice of deception. All his enemies knew this. One time, he was defending a city and the invading army was overwhelming in numbers. He was completely surrounded and the situation seemed hopeless. So he ordered all his soldiers in the city to hide and make themselves invisible. Then he ordered the guards to open wide the main gate of the city wall. He placed himself on top of the wall above the open gate and played nonchalantly his lute. The general of the invading army brought his invading horde to halt in front of the open gate and listened to Zhu-Kuo. It was obvious that Zhu-Kuo was either defenseless or he had laid a trap. But which was it? After due consideration, he considered it better part of valor to withdraw his army. Thus Zhu-Kuo saved the city and won again! This story is well known not only in China, but also in Japan, Korea and Vietnam, where the influence of Chinese culture is strong. Operas and poems have been written about it. It is called "Stratagem of the Empty City".

Zhu-Kuo was popular, not only because of his tricks, but because he was respected. He was considered the epitome of the moral man. He was so powerful that he could have easily become the overlord. Instead he remained true to his chosen lord, Liu Bei, and after his death, to his successor, his imbecile son. His devotion was made famous by the saying in Chinese alluding to him that "he consecrated everything to his duty until his death". Sure enough, Zhu-Kuo died during his last military campaign to save the Zhu Kingdom. Soon afterwards, his kingdom was destroyed by the other two kingdoms. This type of loyalty was much admired in China. It is Chinese chivalry, and may be considered an integral part of Chinese culture. It has influenced my father and me.

Egypt and Mexico were the two most important countries where father was Ambassador. He was Ambassador to Egypt from 1947 to 1956, and from 1958 to 1964, he was Ambassador to Mexico. Thus, he was in these two countries for a total of seventeen years. Egypt is a country which is at the crossroads between Asia and Europe. His nine years there began with last years of King Farouk (1920–1965), who was overthrown in a bloodless revolution followed by the rule of Mohammed Naguib (1901–1984) and Gamal Abdel Nasser (1918–1970). At that time, the situation in China was like one of change of "dynasties" of old. While Father was in Egypt, the national government was driven from the mainland, and in 1949, the People's Republic of China was established. The Nationalist government found a refuge in Taiwan. During such times of change and turmoil in Chinese history, it was not unusual for many of the old guard to switch to the new. But it is typical of my father that he remained steadfast in his loyalty to the Nationalist government.

At that time, Generalissimo Chiang Kai Shek had retired temporarily and Li Tsung-jen assumed the presidency and conducted negotiations with the Communists. There was a movement by some in the foreign service to switch to the Communist side. Such a group had formed itself in Paris and they called upon all members of the foreign service to give up the Nationalist government and switch to the Communists. My father instead did exactly the reverse, and was busily involved in printing circulars advocating loyalty to the Nationalist government. He was also the one who telegraphed Generalissimo Chiang asking him to resume the presidency. Chiang replied, and suggested that this might soon be in the offing. My father's loyalty to the Nationalists is undoubtedly influenced by his loyalty to Chiang as an individual. This is traditional Chinese loyalty, like that of Zhu-Kuo Liang. It is an expression of Confucian ethics.

After Chiang left the mainland and assumed the presidency again in Taiwan, the situation actually improved. At least members of the foreign service were reassured of their continued employment and were paid, whereas their pay had stopped temporarily under President Li. But my father's job became increasingly difficult. He was at the frontline of trying to maintain diplomatic relations in his accredited country. In the beginning, by extraordinary efforts, he maintained the diplomatic recognition of the Egyptian government. Naguib was a personal friend of Father and he was favorably inclined towards the Nationalists. However after Nassar took over, the situation got from bad to worse. Powerful foreign countries, such as India, under Nehru, and Great Britain under foreign minister Bevin, were sympathetic to communist China, and maneuvered for Egypt's recognition of communist China. The inevitable could not be avoided and Father had to at last give up the ship in 1956. His efforts in Egypt were much appreciated by the Nationalist government. The old Generalissimo invited him to dinner upon his return to Taiwan. And at dinner, he repeatedly instructed his son, Chiang Chin-Kuo, the subsequent President, to meet with my father to discuss matters with him. But this conversation never took place. While in Taiwan, Father voluntarily participated in a training program for high officials. This was sort of a semi-military camp for redoctrination. Though he was over 50 years old, he joined the younger ones, and as was his usual habit, he studied hard and performed well. At graduation, he was number one among the 500 or so participants.

The first challenge of being envoy to Mexico, was that he was placed in a different culture with a foreign language. Arabic is spoken in Egypt, but he like other foreigners did not bother to learn the language. They got along either in English or in French. However in Mexico, if you are not able to use the Spanish language, you are handicapped. So as soon as Father arrived in Mexico City, he engaged a teacher of Spanish and started to learn the language. Spanish-speaking people are somewhat like the French; they are proud of their language and would rather not use English, even if they know it. After one or two years in Mexico, father was able to speak and

Monto and Carol Ho visiting Monto's parents, Ambassador Feng-Shan Ho and Mrs Shiao-Yun Ho at the Chinese Embassy, Mexico City, 1959.

write Spanish. His lifelong habit of attacking a task with full vigor did him well.

In order to prevent Mexico from recognizing the communist regime, like in Egypt, he vigorously engaged in a number of activities. He negotiated a commercial treaty between Mexico and Nationalist China. He did his best to prevent the cultural and commercial visits from mainland China. Many years later, I met a correspondent from Taiwan who had been stationed in Beijing and knew the people in their Ministry of Foreign Affairs. They, on the Communist side, reminisced about father's work in Egypt and in Mexico, and noted his relentless efforts to maintain the recognition of Nationalist China, and they expressed admiration for his efforts. Those in the know realized what he tried to do, even though the ultimate outcome was irrevocable. The recognition of communist China eventually took place.

Before going to Israel in 2001 to get Father's award, I did not understand what Yad Vashem meant. It is many things. It has a historical museum complex, containing extensive memorabilia of the sufferers, and the world's largest and most comprehensive repository of materials on the Holocaust. In 1953 Yad Vashem was authorized by the Israeli parliament (Knesset) to bestow the title of "Righteous among the Nations" to thank and honor individuals who risked their lives to help Jews during the Holocaust. Documents supporting nominees are thoroughly researched and validated by the "Committee for the Designation of the Righteous", chaired by a supreme court judge. By January 1, 2001, 18,269 individuals of 35 nations were so honored[3]. Poland (5,503), Netherlands (4,376), France (2,008), Ukraine (1,609) and Belgium (1,247) were the nations with most honorees. They were the nations with over a thousand individuals. Among the honorees, only 20 were diplomats. One of the best known is Raoul Wallenberg of Sweden. He was a diplomat stationed in Budapest, Hungary, where he saved thousands of Jews from being transported to concentration camps and death before the end of World War II. There is only one other Asian diplomat besides Father in the group, Chiune Sugihara, the Japanese Consul in Kaunas, Lithuania. Like my father, he issued visas in violation of his government's orders.

The weather on January 23 seemed to share in the excitement of the day. It alternated a fine drizzle with somber clouds and bright dazzling sunshine. There was an immense crowd, including reporters from all over the world, in front of the "Hall of Remembrance" of Yad Vashem. I was asked by an Associated Press reporter to make a statement in both Chinese and English. There were reporters from both the People's Republic of China and Taiwan.

There was first a ceremony in the "Hall of Remembrance". Manli and I lit a flame from the eternal flame. A cantor canted soulfully in Hebrew. Then the audience moved to the adjoining auditorium. All the seats were soon occupied. What struck me was that there were many high school boys and girls. Apparently, they are encouraged to go to such events in Israel. Mr. Paldiel, the director of the "Righteous among the Nations Department" of Yad Vashem, presided. I got to know Mr. Paldiel fairly well during my few days in Jerusalem. He is an intelligent, cosmopolitan and knowledgeable gentleman with an impressive familiarity of different languages. He translated everything said during the assembly from English to Hebrew and vise versa.

Mr. Maltz, chair of the evaluating committee recounted the process of evaluating documents relating to Father. There was "never a question" about his eligibility. Then Mr. Paldiel awarded the medal and two certificates to Manli and me as we both stood on the platform. This was followed by a speech by Manli and by me.

Manli's speech was highly polished and beautifully delivered. Her speech was interspersed with quotations and Hebrew phrases, which went over well with the audience. Why did it take over sixty years for Father's work to be recognized? What he did for the Jews was not popular at the time. He did not broadcast his deeds. He never imagined that he would be rewarded for what he did; certainly not after he was dead. Because he had violated his superior's order, he did not record in detail what he did, even in his book. He did not record, for example, how many visas were issued. After his death in 1997, Manli, who was trained as a newspaper reporter, wrote his obituary, which was published in the international press. In it was an item which mentioned that he had been the Chinese Consul-General in Vienna and that he had helped the Jews. This information interested an Eric Saul of San Francisco. Mr. Saul is an American Jewish businessman whose interest is to seek out and research non-Jewish diplomats who helped the Jews during the Holocaust. He was instrumental in discovering Chiune Sugihara of Japan. He began to collaborate with Manli, and together, during a three-year period, they discovered many pieces of evidence of Father's deeds, among surviving Jews or their surviving relatives throughout the world. They found out by comparing the serial numbers of their visas, and from documents of the Ministry of Foreign Affairs which I obtained in Taiwan, that the minimal number of visas issued was at least one thousand. From 1998 to 2000, they gathered Father's materials and made exhibits of his exploits, which were shown in Jerusalem, Stockholm, Geneva, Toronto, New London, Seattle, and the United Nations in New York. They submitted their documentation of Father's deeds on August 7, 2000 to Yad Vashem for evaluation. The award was announced by Yad Vashem in October.

I gave a shorter speech, extemporaneously. In the beginning I saluted the "boys and girls" in the audience. I told about being with Father in Vienna from 1937 to 1940, when I was ten years old. I described Father as an active, popular diplomat, fluent in German. I described the German annexation (Anschluss) of Austria. I watched German troops parade the streets. I felt the enthusiasm of the man in the street. But for the Jews, the handwriting was on the wall. They sought ways to leave greater Germany. Father's visas proved to be a lease on life. This event is now being recognized by Yad Vashem. In the last analysis, it is what happens to individuals and what individuals do that matters. There are details of individuals suffering and dying during the Holocaust. These are exihibited and studied at Yad Vashem. Individuals perpetrated the crimes of the Holocaust. In the aftermath, Jews knew how to find them and convict them. Some gentile individuals rescued the persecuted. There are not many such individuals, considering the six million that were killed. But Yad Vashem knows how to seek them out one by one. Precise documentation and acknowledgement of these three aspects of individual events and actions is what the work of Yad Vashem is about. This precise and relentless approach of the Jewish people to crimes against humanity is what other nations need to emulate.

The last person to speak at the ceremony was Frida Rogel, who went from Vienna to Shanghai at the age of seven. She spoke for the Jews that Father helped.

After the speeches, we went to view the collection of stone monuments on which the names of the "honored among the nations" are carved. Previously, each honoree had a tree planted for him or her. But space soon ran out for the multitude of trees. Now, every country with honorees had a stone monument. For China, there is a monument with only two names on it; Pan-Jun Shun and now my father. Shun was a Chinese immigrant to Ukraine, where he protected a Jewish girl from the Germans during World War II.

From left to right: John Ho, Bettie Ho Carlson, Feng-Shan Ho, Shiao-Yun Ho, Carol Tsu Ho, Monto Ho at our Pittsburgh home, 1985.

Father was also honored that same year, 2001, in China. My father was born on September 8, 1901. His birthday in 2001 was the centennial of his birth, and it was celebrated in China. Most of our family — that is, Manli and her husband John Wood, my wife Carol, our daughter Bettie Pei-wen, her husband Ron Carlson, and their twin daughters, 8-year-old Caroline and Margaret, went together to Yiyang, Hunan, Father's birthplace, for a celebration sponsored by the municipal government of Yiyang. His being honored in Israel was also an important part of the celebration. They arranged a celebration of festive activities for three days. One day there was an assembly of about 500 people. There were speeches from guests from China, Israel and the United States. Manli and I spoke again. I gave my speech in both Chinese and English, so that our American relatives, including our granddaughters, could understand it. Two exhibits of over a hundred pieces were shown in Yiyang and in Changsha, the capital of the province of Hunan. They were brought from the United States by Manli. The exhibits consisted of pictures of Father and his family in different countries where he was stationed. The exhibits were subsequently also seen in Beijing and Shanghai. These events received widespread news coverage.

The parents of the Secretary General of the World Jewish Congress and Vice Chairman of the Yad Vashem Council, Dr. Israel Singer, traveled to Cuba with the visa from Father's consulate.

Mr. Singer said, "My parents were saved by Dr. Ho. He is a true hero and I want to introduce him to the world."

part

| II |

ADULTHOOD AND CAREER

| 5 |

FROM PHILOSOPHY TO MEDICINE

I t was at the end of my sophomore year at Tsinghua University, and I was saying goodbye to my schoolmates before leaving for Harvard College in the United States. For a Chinese undergraduate college student, to go to study in the United States in 1947 was a privilege and a challenge. The challenge was whether I would do well. According to the Chinese way of looking at things, to study is an opportunity for improving oneself, not only in terms of book knowledge, but morally. It was character building. When one is given the opportunity to study abroad, expectations are accordingly higher. These were the thoughts of my schoolmates and me as they wrote in my memory book and I said good-bye to them, one by one. When I came to Chang Kuo-chao, my roommate and classmate at Nankai, Lienta and Tsinghua, he wrote:

> "I will be meeting your boat at the dock on your return to China, and I expect that by then you would have realized some of the dreams that you and I have dreamt together."

When my father was appointed by the government to be the Chinese Ambassador to Egypt in the spring of 1947, he was permitted by the regulations of the Ministry of Foreign Affairs to acquire passports for his children to go abroad with him. At that time, the only usual way for Chinese students to go abroad was to graduate first from college and then to take the examinations to study abroad, either for a government scholarship or by self pay. This system did not permit many to go abroad, but to go abroad before graduating was unheard of. When Father received his new appointment, he asked me whether I would like to go to the United States to study.

My response was that I was really quite happy at Tsinghua University, and I did not have a strong desire to go abroad. Perhaps I was a bit afraid to be so different from my peers, none of whom could even dream of going abroad. However, on thinking the matter over, it became apparent to me that I was being offered a very unusual opportunity. The standards of the best universities in the United States were the best in the world. I had to admit that Tsinghua could not compete. Furthermore, in the six years that I had spent in Chinese schools since my return to China from the United States in 1941, I had basically accomplished my primary goals. I had become literate in the Chinese language. I had acquired the mannerisms, likes and dislikes, and aspirations and ambitions of a Chinese college student. I had qualified as a Chinese patriot by being a good student. I had become thoroughly Chinese. There was really nothing more I could learn by remaining in China. I decided to accept my father's proposal by applying to a single American university, Harvard. I would have remained in Tsinghua were I not accepted. I knew little about Harvard at that time and did not attempt to find out more about it. All I knew was that it was the most prestigious university in the United States, and that was good enough for me. One month after I applied, Harvard University sent me a letter informing me that I had been accepted in transfer, and I was asked to register in September. I told my father and he was happy.

Chang was not the robust, athletic type. He was slim and had narrow shoulders. He had a big head, and his nickname was "Big Head". His face was full, kind and benign. His voice was resonant and pleasant. Quick on the draw in conversation, he was witty and exuded intelligence. The foremost impression of him was his modesty. He studied hard because he thought he was inadequate. But actually he was an outstanding student, excelling in the important subjects, Chinese, English and mathematics. Chang came from a prominent family of accomplished intellectuals, on both his father's and mother's side. His father, Carson Chun-mai Chang, was a famous, somewhat iconoclastic philosophy professor, who was the head of one of the very few political parties outside of the Kuomintang that were allowed to exist under Chiang Kai-shek, the Democratic Socialist Party. He got from his father, who often wrote in the classical style rather than the vernacular in the lay press, his devotion to the Chinese classics. Chang would study the Chinese classic off and on in college, though his major was physics. I never observed this in any other student at Tsinghua.

He and I were good friends, in fact I considered him my best friend. What we shared most avidly was our patriotism. Both of us were deeply concerned about the future of China. And we talked constantly about it. This was a focus for us when we were together at Nankai, Lienta and Tsinghua. Our dream for the future was to do something for China.

After graduation from Nankai Middle School in 1945, Chang and I passed the college entrance examination and got our first choice, which was the Southwest Associated University (Lienta) in Kunming, in faraway Yunan, southwest China. Lienta was an association of three very famous universities in the North, i.e., Beijing and Tsinghua Universities in Beijing and Nankai University in Tianjing. When the war with Japan broke out in 1937, these three schools united and moved to the Southwest. Lienta had a very strong faculty, as it drew upon the faculties of all three universities.

Life in Chinese universities was very different from life in secondary schools. There was essentially no contact between boys and girls at Nankai Middle School. There were few parietal rules concerning behavior in college. There was much greater freedom between the sexes. The physical buildings and housing were very poor at the University. The dormitories were essentially a group of bungalow shacks with straw roofs. On the other hand, living in Kunming was a delight. One reason was the marvelous weather. All seasons were like spring. Even though Kunming was close to the equator, because of its high altitude, it was neither hot nor cold. Most remarkably, the skies were usually azure blue and cloudless. The air was fresh, clean and devoid of pollution. These characteristics made Kunming different from most cities in China. Even though most students were poor, they did not lack for the basics. Tuition was free, and most of us were even given a modest government allowance. We did not lack clothing and food, which was better than at Nankai.

I first majored in chemistry at Lienta. My interest in chemistry was stimulated at Nankai Middle School by Chen Xin-Ting, our chemistry teacher. He was a good looking, somewhat stocky man of about 50 years of age. He was Caucasoid in appearance, with a ruddy face and a prominent nose. He was an old timer who came from the original Nankai Middle School in Tianjing. He spoke Mandarin with a heavy Tianjing accent, which because of its accent of the lower tones, gave his speech a not unpleasant flat inflection. He was the only teacher in the entire school who would arrive at each class dressed in a long, white, spotless and ironed laboratory coat. He was very well grounded in chemistry and was careful and meticulous in his teaching. His notes were in English. They were carefully printed on the blackboard in block letters for us to copy. They were crystal clear, even though he pronounced English atrociously. Besides lectures, he also conducted experiments with us. His course was really college level since we had already had chemistry in junior middle school. Teacher Chen was admired and appreciated by his students. As the result of his influence, many of my classmates majored in chemistry or chemical engineering in college. I was one of them.

A chemistry major at Lienta had as freshmen three courses in chemistry: inorganic chemistry, quantitative analysis and qualitative analysis laboratory. These courses were specifically for majors. Our experiments in the laboratory course had to be conducted outdoors, because the fire hazard was too great inside the poor buildings. Two of our three instructors were well known chemistry professors. Zhang Qing-lien received his doctorate in Sweden, where he did research on heavy water. He taught us inorganic chemistry and was personally interested in my progress. I was interested in my courses and did very well in them. I felt particularly happy in doing well in quantitative analysis, where my instructor was Professor Chiu Ching-chuen. This course involved analytic thinking and calculations. I got the best grade in class. My demonstrated aptitude in this aspect of science gave me confidence that I could do the work, when I eventually chose medicine as a career after I went to the United States.

1945 was a momentous year for the world. After six years of total war, World War II had ended in unconditional victory for the Allies, first against Germany, and then against Japan. China, as one of the "Big Four", had finally overcome one hundred years of humiliation and suppression by the western powers and Japan. Japanese troops which occupied most of her coast and much of her inland territories, had surrendered to Chinese commanders and were repatriated to Japan. Taiwan was formally returned to China. The western powers, primarily the United Kingdom and the United States, had abrogated unequal and discriminatory treaties against her. The international settlements in her cities were gone. The American exclusion laws against the Chinese were annulled. China was one of the five permanent members of the security council of the United Nations. I, who had returned to China in 1941 because I was patriotic and wanted to help her, should have been ecstatic. The fact however was that, even as the war was heading toward inevitable victory, a great malaise palled over China. The Kuomintang government of Chiang Kai-Shek was gradually losing the confidence of the Chinese people. It was in the throes of a life and death struggle with the Communists. The Communists, who were a minuscule minority at the beginning of the war in 1937, had expanded to the point of occupying about a third of the entire country. What was more important was that while poverty, chaos, inefficiency and corruption was rampant in areas ruled by the Kuomintang, the Communists were coming through as being austere, orderly, efficient and honest.

It is a bit ironical that the Communists used "democracy" as a powerful weapon against the Nationalists. They supported, if not instigated, a mass propaganda line in which the majority of the non-committed public, especially the students and intellectuals, were eventually estranged from the regime of Chiang Kai-shek. In this struggle, my father and I differed. My father remained faithful to the regime. I, however, went through a long period of doubt and skepticism.

Besides academic excellence, Lienta carried on the great tradition of Beijing University, where the great May-Fourth Movement of 1919 originated. One felt much freer at Lienta, since Kunming was far away from the war time capital. There was among the students and faculty an amazing degree of open resentment against the Kuomintang regime, whose capital was in Chungking. There was an intense interest in the politics of the day. Many political groups were active. The Communists even published a newspaper. Some at Lienta joined the Communist Party, which we later heard was strong in some segments. I knew none at the time among my own acquaintances. The atmosphere at Lienta was much influenced by the assassination of a famous professor, a poet named Wen Yi-to. His death was attributed to the Kuomintang security forces of the government. This incident was used to promote demonstrations and a student strike. The group of students Chang and I associated with were concerned about national affairs and the future of the country, but did not participate in actual political activities. We wanted to find out what the real political and economic problems were and how they could be addressed. After our discussions we would frequently seek out original literature or articles to read and study.

What I did was to intellectualize the problem of the day. China seemed to be at a crossroad. What was the best system for China: Marxism, liberal democracy or something in between? I thought the proper answer would come with proper knowledge and analysis.

My interest in politics became the turning point in my academic pursuits in college. I decided to transfer to the department of political science the next school year (1946) when we moved from Kunming to Beijing. I did this, not because I was not interested in chemistry, but because I thought I had a more immediate problem to solve. My teacher, Professor Zhang of the chemistry department, publicly regretted my decision to transfer. My quest for answers in politics was continued after I went to Harvard College to study.

Lienta was dissolved in 1946, and the original three universities which made it up returned to their original sites in the North. We could choose which university we wanted to attend. Chang and I chose Tsinghua University in Beijing.

Beijing in 1946 was a most interesting city. One thing that distinguished it from all other Chinese cities I knew of was the politeness and courtesy of its people. People from other Chinese cities were frequently rude and impolite, at least to strangers. The ordinary man in the street in Beijing, the laborers, shopkeepers or pedicab drivers, were invariably polite and considerate. The hallmark of this was the extreme courtesy of their everyday speech. They had idiosyncratic expressions which were very pleasing to the ear. Whenever they talked, every few sentences seemed to be interspersed with idioms equivalent to "excuse me" or "please" or "may I borrow the way", etc. These expressions were not known in any other part of China, as far as I was aware. It may have been part of the old Beijing dialect which is erroneously called "Mandarin". (Mandarin in English is really the universal national dialect of the Chinese, which is close to the Beijing dialect, but not identical.) In addition to its attractive people and language, Beijing was a truly beautiful city, which rightfully was the historical capital of China. Looked at from on high, it was lush and green, with the Dagoba, a globular, white Buddhist temple, sitting in the middle. Fifty years later, I returned to Beijing to recall these early impressions. My previous memories of the place and its people seem to have completely disappeared. It is difficult

Monto skating at Tsinghua University, Beijing, 1946.

in today's Beijing even to hear the authentic Beijing dialect. As far as the city is concerned, it has become a megalopolis, with increasing numbers of skyscrapers, but I could not find the tranquil charm of old Beijing.

Tsinghua University was situated in the western suburb of Beijing, in a place called "Tsinghua Yuan", which was an estate of a previous Manchu aristocrat. Its extensive grounds were heavily wooded, interspersed with many ponds and beautiful walks. The physical structures of the university were damaged during the Japanese occupation. Their troops used the university for barracks. The gymnasium was used as stables for their horses. But despite destruction, in comparison with the physical environment of Lienta, Tsinghua was a vast improvement. At Tsinghua I started a different style of studying. As I was now majoring in the social sciences and humanities, I thought it was important to seek out as frequently as possible the original sources. My basic principle was that, besides learning from lectures heard in classrooms, I would fortify myself with knowledge acquired through reading. I would frequently go to the library and look up Western sources in history and philosophy. I was helped by the fact that I could read English and German, but the capacity to read technical material in these two languages had to be developed at the time. I did not have it coming from Europe and the United States. It was acquired by self study in the library. At that time, very few students used the library at all, except to study. I was often the only person wandering among its stacks. Chinese students in colleges and universities have the custom of relying primarily on the lectures of the professors and on their notes to study. It is not possible to delve deeply into the social or humanistic sciences relying on these sources alone.

Chang and I were suffused with youthful idealism. Some of our idealism was abiding and lasting, and it played a significant role in our later lives. In Chang's case, patriotism and idealism led to his becoming a political activist after the Communist takeover in 1949. He was caught up in the euphoria that the Communists engendered, which blinded some patriots after they took over China in 1949. He joined enthusiastically in their movements, while others were more cautious and stayed away. He volunteered in the people's "volunteer" army as a political worker, when the Chinese Communists decided to intervene in the war in Korea. He was stationed in northeast China, near the Korean border, where he died in 1951. He was said to have committed suicide.

By coming to the United States in 1947, I avoided the fate of Chang Kuo-chao. I did not have to choose being embroiled in the maelstrom of

Monto at Harvard College, bridge across Charles River, Leverett House; Monto's house in background, 1947.

political activism. I pursued the scholarly path of expressing my patriotism. The patriotism that Chang and I shared led me instead to study political theory at Harvard, and to a major in philosophy and government. That was the outcome of my transfer from chemistry to political science at Tsinghua. It culminated in my college thesis at Harvard College, an intellectual tour de force in which I solved for myself, or at least reached an impasse concerning the question of "what is the best political system for China?".

After arriving in Boston in 1947, I went to Harvard College to register. The simplicity of the procedure astounded me. The lady who helped me to register was a senior secretary. She looked at my transcript of two years of college work at Southwest Associated University and Tsinghua University. Then she read the letters in English which I had written. Finally, she asked me why was it that I was only transferring to the second year class and not the third year. I said that probably Harvard thought I needed some time to catch up, and perhaps I needed help in English. She said that the credits I had already accumulated in the two years of college in China were more than enough to permit me to transfer to the junior class. As for my English, she said that according to what she observed after conversing with me, and after reading over my letters, she thought that I had no problem with the language. She therefore decided that I should transfer to the junior class, and that was that. The whole process would be inconceivable in China, where every such decision had to be checked and rechecked and substantiated by documents. It would have taken days to arrive at the decisions that took her thirty minutes to make. This gave me a profound impression of how Americans do things. They work efficiently and deal directly with the problem, with a minimum of bureaucratic encumbrances.

I looked through their pamphlet describing the major areas of concentration for college students, and found in it an area of concentration called "philosophy and government". This combination seemed to be exactly what I wanted, so I chose it as my "area of concentration" or major, even though there was no other student in it. The usual load for a student per term was four courses. This is equivalent to twelve credits in a usual college or university in China or in the United States. At Tsinghua, we would take at least twenty credits a term. So I felt that twelve credits were not enough and I decided to take five courses or fifteen credits. My advisor, Professor Ware, who had to approve my choice unfortunately did not object. This is because he was a professor of Chinese literature and not very interested or knowledgeable in what I wanted or was doing. The decision was a big mistake. I did not realize that the big difference between Harvard and a Chinese university was that, besides attending class, one was expected to study and read extensively outside of class. And this is the basic reason why only four courses are recommended. One of the courses I selected was the major, advanced, graduate course in the Department of Government called "political theory". It was taught by a famous and demanding professor of German origin, Dr. Carl Friedrich. His was a huge class with more than one hundred students. At the end of the first lecture, he gave us our reading assignment. To my horror, it covered two full blackboards, with more than twenty different items. This was a major revelation to me and I began to understand what was expected of me at Harvard.

What was the best political system for China seemed to be a reasonable question to ask since she was in a civil war and at a crossroad, with several possible alternatives. One of the alternatives was communism, since it was the doctrine of one side of the civil war that was raging in China when I left. One of the attractions about communism was its apparent profundity. It is especially attractive to intellectuals. The message of the short "Communist Manifesto"[4], stirred me when I read it one hundred years later at Harvard. It was a call upon the proletariat of the world to unite, for the "proletariat of the world have nothing to lose but their chains". But this outwardly emotional call was presumably based on a social theory of inevitability. The victory of the proletariat was "scientifically predetermined". Social development, according to Marx, proceeded according to "scientific laws", that were based on "dialectic materialism". These were obscure terms, that were incomprehensible not only to the common man in the street, but probably also to most communists themselves. My motive in majoring in philosophy was to try to understand these terms. In philosophy, Marx was a disciple of Hegel, who was a German idealist and probably one of the most obscurantist of German philosophers of the nineteenth century.

The foundation in philosophy I acquired during my two years at Harvard College was perfectly suited to my purpose. I took most of the courses that were offered, including postgraduate ones. They covered the history of western philosophy, Plato, theory of knowledge and metaphysics. Before graduation I took the graduate record examination in philosophy, and I received almost full marks.

During my senior year, I elected to write an honors dissertation for graduation. One is not ordinarily expected to write a dissertation unless one elects to go for "honors". In order to write this dissertation I first had to choose a suitable topic. It had to satisfy my quest.

At this time I discovered in the library the work of a philosopher called Karl Popper. Two of his works were involved, *The Open Society and its Enemies*[5], and *Logik der Forschung, zur Erkenntnistheorie der Naturwissenschaft (Logic of Research, the Theory of Knowledge of Science)*[6]. The first volume was readily available in the departmental library, as it was a reference for my course in political theory. But I had to dig the second volume out of Widner Library. It was an older work and in German. It was a basic philosophic work. It provided a lucid definition of what is meant by "scientific". This was important to me because without a clear definition, it would not be possible to critique whether something was scientific or not. According to Popper, a scientific law is a hypothesis that can be negated or disproved by empirical data. If the proposition is such that it cannot be negated (such as "there is a God"), or if no available empiric data could negate it, then it is not scientific.

Another major tenet of Popper's was the nature of history. Historical phenomena are not like phenomena in the physical world, which are repetitive and reproducible and can be subject to experimentation. Historical events are unique. Popper denied that laws that predict historical developments could be found. The first volume used these ideas to critique political and social theories. Looking at the whole gamut of social theories of prominent western thinkers and philosophers, many outstanding ones such as those of Plato, Spinoza, Hegel, Marx and even Toynbee's theory of historical development were found to be "essentialist" and "historicist". All of them predicted a "closed" society governed by some central principle, law, idea, philosophy or "taboo". They subsisted on unprovable beliefs. Not uninfluenced by the events of World War II, Popper's critique condemned all authoritarian "isms" of the day. Besides Marxism, all forms of socialism, Nazism, and Fascism were condemned. Popper instead advocated a social science in which social problems would be solved "piecemeal" rather than totally. This approach to the social sciences, he maintained, was the only way consistent with freedom of man. His were powerful critical tools. Not only were grand utopian designs criticized, but selective theories of human behavior based on unproven central, essential ideas became suspect. An important target of attack was Freud's psychoanalysis, which at the time dominated the field of human psychological development and psychiatry. Psychoanalysis was considered unscientific because it was based on central unprovable ideas like the subconscious. More recently, Adolf Grünbaum of the University of Pittsburgh has meticulously expanded Popper's critique of Freud's psychoanalysis by showing that some of Freud's central pychodynamic ideas, such as "repression", were not scientifically proven[7]. Perhaps even more telling has been the absence of controlled clinical trials to attest to the validity of psychoanalytic therapy, more than a hundred years after its founding. These factors account, at least in part, for the intellectual morbid state in which psychoanalysis finds itself today.

The thesis topic I chose was rather grandiose. It was "An Inquiry into Two Theories of Social Science and a Dissertation on their Relevance to the Problems of our Age".

The two theories that I chose to critique were two extremes. One was Popper's view of piecemeal social science and the other, the social science of Karl Mannheim. Mannheim was a contemporary social scientist who believed that all human behavior is socially determined[8]. In fact, he believed that the freest of man's endeavors, knowledge itself, is socially determined. He proposed the study of the "sociology of knowledge"[9]. My object was to use Popper's definition of science to critique these two extremes of social science. My basic question was, what kind of social science is possible? How are we to understand social phenomena, in any sense?

Mannheim believed that the causes of all social phenomena and all aspects of social development were knowable, which constituted the study of sociology. It followed that all aspects of social phenomena and development could be manipulated and planned by man. Although he rejected totalitarian "isms", he felt that eventual total social planning was the inevitable result of advances in sociology. Using Popper's definitions, Mannheim's sociology has all the faults of theories of closed society. They depended on unproven central theories, or "principia media" in Mannheim's case. His sociology was not scientific, according to Popper's criteria. Another way of looking at it is that society is so complex that the type of sociology he envisions may not be possible.

My criticism of Popper's piecemeal sociology was that it too may be criticized by his own criteria. In order to explain any phenomenon it is necessary to have some theory or hypothesis. Unless this theory can be empirically negated, it is by definition not scientific. Many theories are used in the social sciences, such as in economics, but they have not been shown to be empirically refutable. Social action of any sort, piecemeal or otherwise, must have some guiding principle. Any guiding principle may be suspected of being "essential" or historicist. The reason is in the nature of social phenomena. They are "historic", that is non-repeatable in nature. The best we have in the social sciences are analogies. But analogies must not be pursued too far if one wants to remain scientific. Basically, historical events are linear and non-repeatable. They are all unique. In that context, it is not possible to do any type of rigorous experimentation.

Popper built his concept of social science as part of the struggle of freedom versus determinism. He considered his formulation to be the championing of freedom against deterministic theories of closed society. He dealt a critical blow to all holistic and historicist theories, but in the aftermath, he left the problem of the social sciences unanswered. Carrying his criterion of science to its logical extreme, all social sciences are unscientific. He could not put social sciences on a solid philosophical basis. This being the case, his position as a champion of freedom becomes strained.

Writing this thesis had a profound effect on me. I felt for the first time that I was at the frontier of advancing knowledge. I had done a significant piece of research. I had used my scholarship to come to some definite conclusions. I felt that I had addressed the question I started with. "What is the best political system for China?" I had no answer but I had some definite conclusions. A few months after I turned in my thesis and graduated from Harvard College, the "People's Republic of China" was proclaimed on October 1, 1949. The new political system in China was a *fait accompli*, it was communism. But I concluded the claim that Marxism is scientific was extremely weak, if not downright wrong. I rejected the Marxist utopia, even if I could not establish any type of social science on solid philosophical grounds. I felt I had considered and understood some of the most difficult problems of human thought, such as freedom and determinism.

These problems are not too far removed from the basic problem of man, including God and religion. What I learnt after writing this thesis was that I could approach these grand questions of humanity with humbleness and confidence. This lesson has lasted my lifetime.

This dissertation was my first original work of any length, and in addition I had decided to type it myself. This created for me an unforeseen problem. During my years of being a student I never worked beyond eleven o'clock at night, when I routinely went to bed. However, the typing of this thesis was an exception. I found myself still typing when eleven o'clock arrived the night before the deadline. I finally finished typing the 150 pages of the thesis at 3 a.m. I submitted it in time to my advisor, Dr. Melvin Richter, who was a graduate assistant in the Department of Government. He read through the thesis, and frankly told me that he did not see any particular merit in it.

I was not discouraged, because I did not feel he was interested in political theory. He was not very knowledgeable in theory or philosophy. He was more involved with minutiae of American politics. I did not feel that he understood the emotional energy that underlay my intellectual pursuit. In retrospect, it is likely that because my ability to express myself was not well honed, the points I wished to make were not clear to him. Nevertheless, he forwarded the thesis to the final judge, Professor Louis Hartz. After reading it, he enthusiastically congratulated me on an excellent piece of work and expressed his appreciation of the insights that I offered. He extolled the unique "vitality" of the thesis. On the basis of this dissertation and my good academic record, I graduated from Harvard College with the bachelor of arts degree, "*magna cum laude*" (high honors).

On rereading this thesis after all these years, I feel that it exhibited an aptitude that was to be put to use in my later life, even though I had given up political theory. The question I posed in the thesis was approached in a frontal, global manner, and my analysis was logical and rigorous. I would later apply the same analytic ability to scientific problems, specifically problems in medical research as documented in two hundred and eighty scientific papers. In a more informal way, I have also tried to apply this ability in my daily life. While the effect there may be more debatable, I did apply it to solve a practical medical problem in Taiwan, the problem of antibiotic resistance. There too, I grasped the problem in a global way and tried to solve it in a logical, analytic way. I find from my college thesis to medical problem solving, there has been a consistency and unity in my basic approach to scientific problems.

In 1948 Popper was a relatively unknown Austrian-British philosopher. It was not until the 1970s and 1980s that Popper became a household name in the philosophy of science, which had become a major specialty of philosophers. His definition of "scientific" has become a classic in the philosophy of science. My "discovery" of Popper is a by product of my excursion in philosophy and social sciences.

One day after my graduation, Carol and I were going by train from Boston to the west coast. Carol Tsu was my girlfriend, who was introduced by our mutual friend Chang Kuo-chao, She had graduated from preparatory school at Dana Hall in Wellesley, where we had met, and went to Scripps College in Claremont, California. To be near her, I elected to undertake graduate work after I graduated from Harvard College in the Department of Political Science of Stanford University with the intention of studying for my graduate degrees.

She suddenly asked, "What is your profession? What do you intend to do in your life?"

This question surprised me. Wasn't it clear from what I was doing what my profession was? As I thought more about it, I realized it was true that it was purely accidental that I chose to go to Stanford, and not to the philosophy department at Yale where I was also accepted. Clearly, I had not decided in my mind whether I was a philosopher or a political scientist. In fact, I had never seriously considered either field to be my profession. My connection with these two fields was based on the nature of the problem I wanted to solve in college: "what is the best political system for China?" This quest resulted in my field of concentration at Harvard: "philosophy and government". Now that I had written my thesis, I had really completed the patriotic intellectual objective that I had set for myself. This quest had become quite academic, for the times had changed. The Chinese People's Republic was proclaimed by Mao Zedong. Now that China was ruled by a foreign and, to me, unknown authority, my capacity to contribute to my country in a political way became questionable. If I had thought of the possibility of a political career, it was no longer tenable. The idea of being a professor had loomed vaguely in the back of my mind, but I never considered it an adequate career. I wanted to do more. I wanted to be of service. I had at the time really no concrete plans for my future. Her question set me thinking seriously.

While I was studying philosophy, and particularly after I completed my honors thesis in college, my outlook on life gradually extended beyond the narrow bounds of patriotism. It was going towards being a decent human being and doing good generally. My question now boiled down to what profession would allow me to do that. The defeat of Germany and Japan in World War II showed me the bankruptcy of militant nationalism. It was no longer tenable in this shrinking world. I was reminded of someone I met after arriving in the United States in 1947. At that time, in order to get from San Francisco to Boston, it was necessary to take a train, which took six days and five nights. Most people spent the nights sitting up in the couch car. The train crossed three thousand miles of the continental United States. Many people got on and off the train on the way; therefore I had the opportunity to meet with and converse with all kinds of people. One particular person came aboard somewhere in Iowa, when we were crossing the great American plains, the rich agricultural country of the United States. He was a middle aged farmer who sat by my side and we started conversing. He was totally unlike any farmer I knew in China. He seemed to be more a gentleman than a farmer to me. But he said he was a working farmer who owned his own farm. I felt the unique self confidence and self respect that an American can exude. In his conversation, he expressed himself freely and without inhibition, exhibiting "individualism" and "freedom" of expression. When he talked about himself and his family, I could feel that he was very proud of them and what he had done. One would expect that his attitude of life would come through as utter selfishness. Actually, the reverse was the case. Along with self confidence, he manifested a form of altruism that is also uniquely American. The outstanding characteristic of American culture is that individuals are proud; they feel equal to any one else, regardless of station, wealth or profession. Part of this pride entails good will to all men. This good will is concretely expressed toward neighbors, making America a country of friendly communities. But Americans also project good will toward all men. I believe this comes from their respect of others as equals. The merits of American culture are not always apparent at first sight. They are obscured by commercialism, Hollywood and professional sports. It is common for foreigners, particularly foreign intellectuals, to feel superior in the beginning. The ordinary American is not particularly "intellectual". It may take foreigners some time to understand and appreciate American culture. My immersion in American culture as a teenager and university student has resulted in my acquiring the American ideal of good will to all men.

After I completed my thesis and reviewed the efforts of the human intellect to achieve the best possible state, I became aware of how difficult the problem was, and how humans have erred in the pursuit of utopias. I was familiar with the fragility of good intentions of individual human beings as well as groups, and I realized that any profession can lead to doing evil. This is true, not only of practitioners of philosophy, social sciences, politics and religion, but even of those in engineering and natural sciences. One has only to reflect on the ethical conundrums in the development of nuclear weapons. Many of the pioneering nuclear physicists who supported its development came to doubt their wisdom in doing so. I came to the conclusion that the profession which is least likely to do evil is medicine. Medicine has a profound intellectual basis which could provide me unlimited range for development. Medicine was different from other professions, because its sole purpose was to reduce human suffering and to do good for individual human beings and not harm them. It never occurred to me to question my aptitude for medicine, or my lack of earlier motivation. I instinctively felt I could do well because I had proven my aptitude for natural science by having successfully majored in chemistry in my freshman year. Therefore, in answer to Carol's question, I decided on medicine as my profession. This drastic change in my orientation took place strictly on the basis of rational deliberation. I did this despite the fact that I was not yet adequately prepared for medicine. Although a graduate student, I was not even qualified as a "premed". I told my father my decision. I was so convinced of the correctness of my decision that I did not think it was necessary to consult him first. This may also be considered as a mark of my independence. It was only years later that I learned that he would have appreciated my consulting him. He never told me directly. I was moved by his tact.

After I finished my medical training in 1954, I thought that I was in a field that was non-political, and that I could return to China and be of service irrespective of my political leanings. I thought I could still fulfill the dream that Chang and I had together, without being entangled by politics. During my post-graduate medical training, when I was a resident at the Boston City Hospital and a fellow with Ed Kass and John Enders, many patriotic Chinese graduate students and young faculty members returned to Communist China. Those were brief, heady days in 1955–1957, after the establishment of the People's Republic of China, when many thought that the millennium had arrived in China. China was at peace, even though the antagonism with the United States was getting worse. Reconstruction of the entire country was forging ahead; and what was more, Mao Zedung had just proclaimed a movement called "let a hundred flowers bloom", designed to encourage freedom of speech among the intellectuals. Carol and I arrived tentatively at the conclusion that we should return after my fellowship with Enders in 1959, and we so informed our parents. Indeed, our preparations had gotten to the point where Carol's mother bought for us woolen underwear, a bicycle and a manual sewing machine and stored them in Hong Kong for us to bring into China for our future Spartan lives there. Despite our tentative decision, many restraining forces held us back. I was surprised to be visited by my stepmother, Huang Shao-yun, in the spring of 1956 in my office at the Boston City Hospital, where I was an assistant resident in medicine. She had, on her own, taken a taxi from Cambridge where they were staying temporarily for a month on their way from Mexico to Taiwan. She tearfully remonstrated with me to withdraw our decision to leave. She recalled Father's bitter experiences with the Chinese Communists in 1927 in Hunan, the memory of which remained with him throughout his life, and is one of the reasons for his steadfast anti-Communism (see Chapter 4). Remarkably, during this crisis of our decision, Father never discussed our decision with us openly. I feel that his insight was such that he probably anticipated that because of my independence of thought, I would not have accepted his arguments. In retrospect, his silence spoke louder than words.

Before we made our final decision, Carol and I decided to visit her cousin, Mary Tu, and her husband, Sam, at Cornell University in Ithaca, New York, in the spring of 1956. The two of us arrived by bus from Boston. They were leaving in a few months for China. Sam was an upcoming assistant professor of electrical engineering at Cornell. After returning to China, Sam had become one of the pioneering architects of China's space program, one of the very few returnees known to us who was left alone to develop in his career without political obstruction. We discussed in Ithaca our decision to return. We agreed that after they returned and observed the prospects, if they agreed with our decision to return, they would write to us. We waited patiently after their return for their word. We did not hear from them. Twenty years later, when we had a joyful reunion in Beijing during one of our visits, they were still silent. Despite their having done so well, I feel that they were ambivalent about their original decision to return to China.

In the meantime, other disturbing events followed in China, one after the other. The "anti-rightist movement" soon followed "let a hundred flowers bloom". Then, in the late 1950s, came the disastrous "Great Leap Forward", when Mao Zedung almost ruined the entire agricultural system of China by collectivizing its farms and forcing everyone to "manufacture steel".

China was brought to the brink of collapse. Then, in 1965–1975, came the infamous "Cultural Revolution", which, during a span of ten long years, all but annihilated China's intellectuals.

We decided to stay in the United States when the time came in 1959. We became immigrants of the United States, and eventually citizens.

I never returned to China the way Chang and I thought I would. I had left for good. Even when I did visit after China was "opened up" by Nixon and Kissinger in 1972, it was by plane and not by boat. Boats had gone out of date. Chang was not there to meet me at the airport. He had died in the 1950s, during the Korean War when he was with the Chinese army in northeast China. What happened? Many in our class at Nankai were shocked and despondent about Chang's fate. In 2000, Zhang Chung-yeu, another one of my classmates from Nankai, Lienta and Tsinghua, went to the relevant government bureaus and investigated this whole affair. He plowed through the red tape to get to the pertinent documents and discovered Chang was denounced because they had found in his pocket a letter from his younger sister in the United States. She had gone there to escape the Communist takeover. Chang committed suicide when he was denounced for being a pro-American intellectual of "capitalist" origin. What mental anguish was engendered when the rationalism and idealism of a highly sensitive intellect confronted the stark reality and irrational anti-Americanism of the Communist army!? Exactly how his denouncement resulted in self destruction is unknown. Zhang Chung-yeu initiated the process of "rehabilitation", a process accepted by the government to redress the wrongs committed during the Cultural Revolution and earlier days of the Communist rule. He was successful. Chang was rehabilitated. His denouncement was formally retracted by the government and his honor was restored; a small but essential consolation in a major tragedy. We all thank Zhang for his work. It is a reflection of the depth of camaraderie among old schoolmates in China.

When Zhang's work became known among Nankai classmates and China had more or less recovered from her woes, Hu Xiao-chi, another classmate, from Nankai Middle School, who remained in China throughout, wrote to me after she read my book in Chinese in 2002[1]. She said, "I am glad that you emigrated from China when you did, and did not return. Otherwise you would not have been able to accomplish as much as you did."

Once the decision was made that I should go into medicine, I acted quickly and decisively. The time was already late in the fall of 1949, and I was matriculated in the graduate program of political science at Stanford. If I wanted to attend medical school the next year, I had to find a school that would accept my application. I discovered that the application deadlines for most medical schools had already passed. But fortunately, Stanford Medical School was still accepting applications. Therefore I applied only to Stanford Medical School. The situation was almost exactly the same three years earlier when I applied only to Harvard College. The medical profession is highly regarded in the United States and the study of medicine is very competitive. It was (and is) notoriously difficult to get into a medical school. There are relatively few schools and they accept only a limited number of students. Prestigious medical schools may take as few as ten percent of their applicants. I considered my chances of being accepted as very low when I applied. In addition, I was not prepared in premedical studies. As a matter of fact, I had never taken a course in biology. In evaluating applicants, besides looking at the record of the applicants, medical schools also depend on personal interviews. During my interview it was apparent that having graduated with high honors from Harvard College was an asset. Still, I was pleasantly surprised when I was accepted. The only proviso was that I had to complete the entire course of premedical studies within a year before I could matriculate.

My medical education began in September, 1950 at Stanford Medical School.

chapter

| 6 |

MY LIFE LONG
COMPANION — CAROL

On June 28, 1952, Carol and I were married in New York. This was the time when she and I left our schools in California and returned to the East. Carol graduated from Scripps College in Claremont, California and I transferred from Stanford Medical School to the third year at Harvard Medical School. We had known each other since 1947.

Early in 1947, I told my best friend, my classmate at Tsinghua University in Beijing, Chang Kuo-chao, that I would be leaving Tsinghua and transferring as a junior to Harvard College in the United States. Chang and I first met when we were both students at Nankai Middle School in Chungking. Before leaving Beijing, he said to me,

> "I must give you Carol Tsu's address. She is a student at a prep school in Wellesley, Massachusetts, close to Harvard. You must go and meet her."

Chang had already made Carol's name familiar to me. He and Carol had become friends in 1943, when both of them were escaping Japanese occupied Shanghai with their families. It was a long and tortuous journey, first through Japanese occupied territory, and then a long wait in occupied Hankow upstream on the Yangtse River. There they waited for an opportunity to slip into free China. Chang and his family succeeded, but Carol and her elder brother, David, failed. They could not wait long enough and had to return to Shanghai. They succeeded the next year, when they joined the rest of their family in Kunming.

I have already written about the common aspirations and patriotism of Chang and me. Of course, young boys like us were interested in other things than idealism. At Nankai Middle School, the boys' and girls' schools were separate but adjacent. Very few of us knew any of the girls. Chang and I knew none until we were both in college. But that did not diminish our interest in them. "Girl watching" was one of our favorite pastimes. We would watch them as they went from class to class, but especially during evenings when boys and girls would take walks on the spacious campus grounds. The appearance of girls were analyzed by us in minutiae. Chang was constantly attracted and infatuated with this or that attractive girl, none of whom he dared to approach.

It was only when we were together in college that Chang told us about Carol. He told me how they got to know each other in Hankow. After separating, they kept up their friendship by correspondence across the Pacific. We were all envious of Chang because he was the only one among us who had a female friend.

He showed me Carol's picture; a teenager with a cute round face, straight hair, twinkling eyes, and a shy smile. He and Carol spent enough time together so that they got to know each other well, like a brother and sister. They were frank and sincere with each other. Chang did not talk about Carol the way he talked about the unknown girls he was attracted to at Nankai. In his inimitable way of pronouncing generalizations, he assessed Carol as being "capable of love".

So when he gave me Carol's address, I felt he was giving me something precious to him; something he would only give to a true friend. Over the years, Carol and I frequently talked about our common friend. We aped his unique mannerisms and his way of speaking. He was such a sincere, intelligent, modest, idealistic, straight-forward and somewhat naïve person; someone whom we admired and loved. We were both devastated by his tragic death in 1951.

In September, 1947, after I had registered at Harvard College and was settled down in my daily routine, I discovered that Carol's preparatory school, Dana Hall in Wellesley, Massachusetts, was only 23 miles northwest of Cambridge. So one day in November, I called her at her school and she agreed to see me. However, we did not realize we had to deal with the very strict regulations of her school. It appeared that female students were not permitted to leave the school to see a boy on their own. She consulted her headmistress, Mrs. Johnston, who was a friend of China, as she had taught English at Yenching University in Beijing. Furthermore, her daughter, Barbara, was Carol's classmate. Mrs. Johnston surprised us all by arranging a cozy and intimate tea for the three of us at her private home. Carol and Barbara had prepared some musical entertainment. Barbara sang and Carol played on the piano Debussy's "Claire de Lune". Afterward, Carol suggested that she and I take a walk. Wellesley is a peaceful, beautiful little town, neat and prim. The day happened to be Thanksgiving Day, and there were very few people walking that afternoon. That was the first day of winter and the first snowflakes were falling on the ground. There was an uncommon stillness felt only at times of snow. I still remember walking in the snow and my footprints imprinted on the fresh snow. This scene has remained with me all these years.

With such a romantic beginning, what could one expect but a successful love story? It seems that there were no obstacles on the way. More than fifty years later, during a celebration of the publishing of my Chinese book in 2002 in Taipei[1], Carol was invited to say a few words. She remarked that her story with me was "love at first sight". Well, I certainly did not feel that was the case then. It turned out that even though Carol was only a high school senior, she had had a number of male acquaintances and her reputation was "hard to get". Fortunately, we had the opportunity to get acquainted further. For two successive summers in 1948 and 1949, we had the unusual opportunity of working at the same place. Without prearrangement, both of us ended up working at Silver Bay on Lake George, New York. We were both employees at a YMCA conference hotel. All employees were college students. They were paid almost nothing, but room and board were provided, and they had access to all the entertainment and sports facilities of the hotel. Silver Bay is truly one of the most beautiful places that I know of. My first impression of the gorgeous lake and its surrounding mountains was that I was literally left breathless.

One should think that under these conditions, we would become rapidly acquainted and befriended. However, after three months we were still like strangers and both of us left Silver Bay that summer without further improvement in our relationship. My impression at the time was that Carol was arrogant and we left without much to say to each other. During the second summer, when we were again at the same place, I met an American girl named Mimi Bowen, and she and I talked confidentially about many things. One day, she remarked to me, "You know, Carol really likes you." Another colleague, Bettie Wei, told me that "Carol is really a 'good' girl". "Good" in Chinese means kind, not arrogant or devious. I was also reminded of Chang's generalization about Carol.

That Carol liked me took me by complete surprise. I was not self-confident enough to realize that I could be liked, or even perhaps loved, by a girl. Thereafter, my confidence increased, and I again approached her, and from then on our relationship advanced without a hitch. At that time, Carol was already a sophomore at Scripps College in Claremont, California, where she had a full tuition and room and board scholarship. After I graduated from Harvard College, I chose to transfer to Stanford University, where I was accepted as a graduate student in the political science department. I transferred from political science to medicine in 1950. When she graduated from college in 1952, I took the opportunity to transfer back to Harvard University. From the day we were married until this day is more than 50 years. Fifty years of marriage is not a trifle! My marriage is one of the great successes of my life.

Our marriage ceremony was entirely arranged by Carol's parents. At that time, my father was the Chinese ambassador in Egypt, and he could not come to my wedding. He was not given a leave of absence. Carol's parents have deep roots in the United States and friends and relatives came to the wedding in large numbers, and for their only daughter they made this an extremely elaborate and festive occasion.

My father-in-law, Y. Y. Tsu, was a bishop of the Chinese Episcopal Church. He wrote an autobiography[10]. He died in 1986 at the age of 101. He was well known during the war, when he was Episcopal bishop of the Yunnan-Kweichow region in China. This is the area where the Burma Road connected China with the outside world. He became well known as "Bishop of the Burma Road" for his many services to the allied military, mostly American. He retired from his work in China and came to the United States in 1952.

Carol's mother, Caroline Huie, was one of the six daughters of a famous Chinese-American pastor in New York, Huie Kin (1854–1934)[11]. Her mother was a Dutch-American, Louise Van Arnam. Because of Caroline's American citizenship, she was able to take Carol and her brothers out of China by flying from Kunming to India, over the "hump", in 1944. From there they landed in Boston via Australia and the Panama Canal, in the midst of World War II. Each of the children went to private schools in eastern United States. David went to Yale College. After graduation he got a master's degree at Brooklyn Polytechnical Institute. He became a nuclear engineer. Robert went to Mount Hermon, and Trinity College in Hartford, Connecticut. Then he went to Virginia Theological Seminary, and became an Episcopal minister. He died in 2004. The youngest boy, Kin, three years younger than Carol, went to Saint Paul's and Princeton College. He got a Ph.D. at Princeton University and became a chemical engineer. He died in 2005.

The nine children of Carol's grandparents were brought up in New York. Interestingly, the six daughters all met Chinese students in New York, married them, and went off to China, while the three boys remained in the United States, married Caucasians and developed their careers in the United States. Carol's aunts, Louise (1893–1991), Alice (1895–1980), Helen (1899–1995), Ruth (1902–1990), and Dorothy (1902–1991) married prominent educators, physicians, and community workers. Louise married Fu-liang Chang (1889–1984), who was a professor at Yale in China College where Father was a student. Alice married Jimmy Yen (1890–1990), who was an internationally known mass educator. He began his work in World War I when he introduced mass education for Chinese workers who were sent to France to work during World War I. From this beginning, he launched world famous mass education movements, first in China and then later, after the Communist takeover, in the Philippines. Helen married Paul Chi-ting Kwei (1895–1961), who was the president of Wuhan University and one of the foremost early physicists of China. Ruth married Henry Hsieh-Chang Chou (1892–1945), who was a dean at Yenching College. And Dorothy married Amos Yi-hui Wong, who was a prominent gynecologist and obstetrician in Shanghai. Nowadays, the offspring of these nine children and their spouses have accumulated in three generations to a clan of two hundred people. Once every three or four years, there is a gigantic family reunion of the "Huie clan" in the United States. This clan is truly one of diversity, as most are Chinese looking, but some are Caucasoid and a few even African-American.

I feel that in order to understand Carol's personality, it is necessary to begin with her family. Her mother and her five sisters, who went to China after their marriages, were especially close. All of them were Christians and all of them had four to six children. Every one of them had a happy family life and marriage. With a family background such as this, it is not necessary to look for abstruse explanations for the ability to be a good wife and mother. All of them are personable and loving people.

Having said as much, I have to emphasize what was unique being with Carol for 50 years. She provided first of all for me the love of a woman, which I sorely missed as a child. Chang perceived the essence of Carol's character. I have been privileged to be loved by her. Secondly, Carol has been consistently supportive of my aspirations, encouraging me in my studies and career. She was the one who first recognized the importance of my choosing a worthwhile profession after college, and she is the one who provided the loving encouragement that allowed me to advance. In addition to that, she has been a peerless mother to our two children, and a loving grandmother to our three grandchildren.

Four years after our marriage, it became apparent that we could not have children. We therefore adopted two — Bettie Pei-wen, born in 1956 and John Chia-wen, born in 1959. These two youngsters came to us when they were five and two months old, and our success in bringing them up to a happy and productive adulthood is another great achievement of our lives.

Bettie Pei-wen was born on September 18, 1956. She graduated from Mt. Lebanon High School in 1974 and from the University of Pittsburgh at Johnstown in 1978. She majored in education. In 1980–1981, she went to Wuhan, China where she taught English at the Wuhan Medical College for a year. That opportunity gave her better knowledge of Chinese culture. She improved her Chinese, which is quite fluent, and she learned to appreciate the richness of friendships among the Chinese. She married Ronald A. Carlson in 1985. Seven years later, she gave birth to delightful twin girls on October 26, 1992. They were named Caroline Louise (Lo-yi) and Margaret Ruth (Lo-tien).

John (Chia-wen) was born on October 17, 1959. He graduated from Mt. Lebanon High School in 1977. A demarcating event in his young life was his obtaining the first prize in biology and the grand prize in the Pittsburgh Region School Science and Engineering Fair in 1975. This was achieved by preparing an exhibition called "The Genetic Transfer in Bacteria." He was helped in developing this exhibit by my friends and colleagues, Drs. Robert B. Yee and Wayne Atchison from my department at the Graduate School of Public Health at the University of Pittsburgh. He graduated from Mt. Lebanon High School with honors in 1977, and from MIT in 1981. He received his M.D. from Brown University in 1985. After a year of internship at the New England Deaconess Hospital in Boston, he attended Amos Tuck Business School at Dartmouth College, and obtained his MBA in 1988. Thereafter, John has had a career in business, specializing as a partner in consulting firms. In 1986, he married Michelle T. Viau. On July 23, 1993, Gregory (Wei-ming) was born.

After the two children had graduated from college, Carol declared that she must have her own independent work and not just be a housewife. She therefore attended the Graduate School of Library and Information Sciences at the University of Pittsburgh, where she graduated with honors and a Master's degree specializing in medical librarianship. This particular choice had the advantage of combining her interests with mine. Fortunately, there are many hospitals in the vicinity of Pittsburgh, and she later chose to become the medical librarian of St. Clair Hospital, close to where we lived. She founded the library in that hospital and took charge of it for 20 years, before she retired with a feeling of fulfillment. Now that she has retired professionally, she continues to be extremely active socially. There are occasions for making many friends among the Chinese in Pittsburgh, as in other large cities in the United States, but in addition to the Chinese, Carol is extremely active among Americans at church and in social and college clubs. She plays bridge, tennis and participates in other activities in the community. Recently, she became Chairperson of the Antiques Group of the South Hills College Club. We are not only lifetime companions, but we also each have our own interests. In retirement, her social know-how among family and friends is a great asset to me as I am more withdrawn.

Carol Tsu, her father Bishop Y. Y. Tsu and her mother Caroline Huie Tsu. Taken in 1947 when Carol was in the last year of Dana Hall Peparatory School, Wellesley, Massachusetts.

Wedding of Monto and Carol Ho, June 28, 1952 in New York City.

Monto and Carol, the newlyweds, Littleton, Massachusetts, 1953.

From left to right: Margaret Carlson, Bettie Pei-wen Carlson, Carol Ho, Ron Carlson, Monto Ho, Caroline Carlson, in Pittsburgh, 2003.

From left to right: Monto Ho, John Chia-wen Ho, Carol Ho, Michelle Viau Ho, and Gregory Wei-ming Ho, in Quito, Ecuador, 2002.

chapter
| 7 |

FELLOWSHIP: FROM ENDOTOXIN TO INTERFERON

One day in 1957 I met with my friend, Robert S. Chang, for lunch around Harvard Medical School and Harvard School of Public Health in Boston. Robert is about five or six years my senior. He was a graduate of St. John's University Medical School in Shanghai. After coming to the United States, he also went the route of infectious diseases and microbiology, and he obtained a doctoral degree at the Harvard School of Public Health. At that time, he was an upcoming assistant professor at Harvard School of Public Health and was becoming known for his work in virology. I was then a research fellow in infectious diseases at the Boston City Hospital. He said,

"You should really consider virology for research because virology has become a thriving part of infectious diseases and microbiology. Virology is an upcoming discipline. Furthermore, one of the greatest virologists, John F. Enders, is right here at Harvard. A few years ago, in 1954, Enders earned the Nobel Prize in medicine for his successful cultivation of the poliovirus in cell cultures. He would be a great one to work for."

I had not considered virology as an area of research, because at the time, traditionally, viral diseases were the domain of pediatricians rather than internists. But it was true that virology had become an important discipline, precisely because of advances that people like Enders had made. Robert piqued my interest, and no sooner said than done, I made an appointment to see Enders. It was so simple to do things like that at Harvard, where

Portrait of John F. Enders, given to Monto Ho, 1959.

there were so many well known medical scientists and it was so easy to meet them.

My impression of Enders the first time we met was that he was a different kind of person. Unlike the typical intense scientific achiever, in a hurry to go on to his next appointment or chore, and who would be anxious to hear what I was about in as few words from me as possible, here was a person who was formally dressed in a leisurely business suit with a bow tie instead of a laboratory coat, and whose demeanor was deliberate, thoughtful and slow. Enders had a strong face, with a square chin and lots of wrinkles. He occasionally put on his half moon spectacles to read something and promptly took them off. He was smoking a pipe, but I never saw any smoke coming out of his pipe. He was stooped over at age sixty, and he seemed quite a bit older. He remained exactly the way he looked until he died in 1985 at the age of 88. He acted as if he had all the time in the world. Our conversation started very informally and we did not immediately talk shop.

Enders came from a wealthy New England family with a home in New London, Connecticut. He went to St. Paul's School and he graduated from Yale University. He got a master's degree in English literature at Harvard and was going on as a doctoral student in philology. One day in 1930, he accompanied his roommate to a lecture by the great Harvard microbiologist, Hans Zinsser, who was well known for his eloquence in his lectures. After hearing him for an hour, Enders made up his mind to give up his doctoral work in English and went to study under Hans Zinsser. He became his doctoral student and then went on to become an academic microbiologist, spending his entire career at Harvard Medical School.

I was struck by the similarity of our backgrounds. Both of us decided on our careers rather late; I at age 23 and he at age 33. Both of us started in college in the humanities, I in philosophy and government, and he in English.

Enders maintained his interest in the humanities as did I. Both of us were interested in the affairs of the world and politics. That was what we talked about in the beginning. He was very well read, especially in the classics and literature. He would quote authors like Oliver Goldsmith, with whom I was not very familiar. Just before he died, he was reading T. S. Eliot to his wife, Carol. After talking with Enders for an hour, I decided that I had to work with him. I seemed to feel the same way as he had with Zinsser. I do not recall what research assignment he gave me, if any. But I felt that irrespective of what research topic he would propose to me or throw my way, I would accept it. He also seemed inclined toward me as he accepted me as a fellow on the spot, with a full time stipend to boot.

It was less than a year after graduating from medical school, and two years of internship and residency in internal medicine that I decided to go on for a fellowship and research. I had become exposed, during my third and fourth years in medical school, to several disease areas. I wanted to understand them better.

I found a book in the library, called *Chinese Lessons to Western Medicine: A Contribution to Geographical Medicine from the Clinics of Peking Union Medical College* by Isidore Snapper[12]. This book gave me a sought for introduction of what medicine was like in China. It satisfied my eagerness and curiosity for knowledge of my native country which I thought was necessary for me to be of service medically among the Chinese. The book was full of cases and descriptions of exotic infectious diseases caused by parasites and microbes, unheard of or unknown in this country. I remember a picture of a young man whose lower chin was destroyed by a "noma", a medical term that I learned from this book. It is gangrenous necrosis, due to destruction of tissue caused by lack of white cells, as a result of enlargement of the spleen. The man had kala azar, a visceral disease caused by a parasitic protozoan, called *Leishmania donovani*. The enlaragement of the spleen is due to the parasite. They are transmitted by sand flies then common in arid northern China. This book ingnited my interest in infectious diseases, which proved to be abiding and lifelong.

Snapper was a visiting professor and chair of the department of medicine between 1940–1942 at the Peking Union Medical College (PUMC), which was a medical school supported by the Rockefeller Foundation[13,14]. It was the best medical college in China, and one of the best in the world. In its 34 years of existence (1917–1951), with a total number of only 311 medical graduates, it had the greatest impact on medicine, public health and research in China of any American-supported institution.

The planning of the school had the indelible imprint of Abraham Flexner (1866–1959)[15], member of the Rockefeller Foundation and author of the famous Flexner Report of 1910, which revolutionized and modernized medical education in the United States[16]. Flexner evaluated each one of the 155 medical schools in the United States and Canada, and considered 132 were below standard. As a result of the general acceptance of the report, only fifty medical schools were left. Many were proprietary schools with inadequate entry requirements which were basically diploma mills. Flexner and his brother, Simon, a bacteriologist, advocated premedical preparation in college before commencing medical school. Both emphasized the importance of basic science in the medical curriculum. Most American medical schools then introduced courses in anatomy, histology, physiology, pharmacology and bacteriology. These courses took the first two years of medical school. In advocating this heavy dose of basic science, they were influenced by the German model. But equal emphasis was given to clinical medicine in the last two years. Flexner's reform emphasized the importance of bedside clinical teaching. In this, he was influenced by the British system.

The Anglo-American tradition of teaching of clinical medicine is different from medical schools on the European continent or in Taiwan. Using Harvard Medical School as an example, it has only 130 students per class. Even the largest medical schools in the United States have only 200 students per class. In order to train this fairly small number of students, Harvard Medical School has four major general teaching hospitals, in addition to two or three specialty hospitals. Why is it necessary to have so many hospitals? The answer is that many patient beds provide an opportunity for hands-on clinical training for the students. The type of hands-on clinical training that the American medical student gets is actually non-existent in many continental medical schools. What they have is not hands-on clinical training, but observational clinical training. Basically, the origin of medical education in Taiwan comes from the European continent through Japan. Therefore, it is not surprising that their attitude towards clinical training for medical students, which I had a chance to observe while I was in Taiwan, is similar to that of the European continent.

A unique characteristic of American clinical training is that it imparts to the student a sense of responsibility for the patient. In the fourth year, where most of the clinical training takes place (equivalent to the sixth year in Taiwan's medical schools), the student clerk on the wards already feels responsible for his patients. He has his assigned position in the hierarchy of care for patients, and he has self-respect in that role. On the other hand, while the number of students in Taiwan's medical classes are now similar to that of the United States, their students are still mostly observers without a sense of direct responsibility. They feel like intruders. In the United States, the responsibility for the care of the patient is divided among a hierarchy of physicians, ranging from the student to the house staff, to the responsible chief resident or "visit". In Taiwan and on the European continent and Japan, all authority seems to be invested in a single person, the almighty professor or the visiting physician.

In the United States, the first person to see a new patient on the medical service is either the fourth year student or the first year resident. This first contact allows the initial observer to make the diagnosis and to propose the initial course of treatment. Admittedly, the final diagnosis and eventual course of treatment must be approved by the senior residents and visiting physicians, but the experience of taking a history and doing a physical on a new patient initially and proposing a diagnosis and course of treatment is invaluable experience for the trainee. He is forced to come up with a meaningful diagnosis, which he has to defend while reporting on the case. On the other hand, while in Taiwan, if I asked the physician in training about a case, his information was usually second hand, reflecting the fact that he did not have initial contact, and he was merely an observer.

My experience in clinical training as a first year intern was at the Harvard Medical Unit of the Boston City Hospital. Each teaching hospital of Harvard Medical School has its own traditions and esprit de corps. At the "City", we were proud of the fact that we were able to do an excellent job despite deficiencies and bureaucratic red tape. One day, I was on call as a night float, assigned to admit patients during the night. Around midnight, I admitted a middle-aged woman who I suspected had hyperthyroidism. Ordinarily, the duty of the night float is to admit the patient, take care of any emergency, and then report the patient to the day staff. I worked up the patient by taking a thorough history, doing a physical examination and then undertaking the basic laboratory procedures, that is examination of the red cell and white cell counts, urine examinations and gram stains of sputum or other exudates. The latter are procedures which physicians nowadays are unfamiliar with, since they no longer do them themselves. After I finished these procedures, it occurred to me that I still had not proven my diagnosis. All I had was a suspicion. Therefore, I scrounged around in the storeroom and found an antique basal metabolism rate (BMR) machine. This outdated but effective machine is designed to measure the consumption of oxygen under basal conditions. At 3 a.m., I got the machine in working order and measured the patient's BMR. Lo and behold, it was elevated and I proved that indeed the patient had hyperthyroidism. Around 7 a.m., I presented the patient to the day staff and they all marveled at the unbelievable completeness of my examination. After such an experience with a patient with hyperthyroidism, is it possible for a first year intern to forget this disease, or not to learn as much as possible about the disease and related diseases? What can be more effective than this kind of clinical training?

It is understandable that after one year of such training, I developed self-confidence and really felt that there was nothing I was not capable of doing. What one acquired was not merely knowledge but experience, which is irreplaceable. When the second year of the residency came around, the nature of training changed again. As assistant resident, I was in charge of a ward with about 30 patients, and I was directly responsible for the fourth year medical student and first year interns. They had initial contact with new patients, whom they followed from the beginning of their stay to the end. I, on the other hand, had to master the conditions of all 30 patients and arrange for their disposal. Each week, we had six mornings during which we had "rounds", which featured the visit of the visiting physician or consultants. I decided who was to be presented and how decisions made at the rounds would be carried out. I had to write a complete report of every patient that was admitted and discharged. There is ample opportunity in the training of American physicians to practice the art of writing and speaking. As a result, after three years of clinical experience, beginning with the fourth year to the second year of residency, I truly felt like a veteran of clinical medicine. In many respects, I believed I had reached at that early stage my apogee in clinical competence.

The new American system of medical education, with all the reforms advocated in the Flexner Report, was incorporated in the curriculum of PUMC.

PUMC sought the best in the world in medicine. This philosophy resonated with the demand for excellence in traditional Chinese scholarship. PUMC brought the best of American professors and it recruited the best Chinese students who were already college graduates. These American professors had already established renown in medicine, surgery, public health and the basic sciences when they were hired, frequently on loan from their parent institutions. The PUMC of the present day, seeks to maintain its tradition of dedication to excellence. But PUMC was still a flash in the pan. PUMC and its American tradition did not become part of the Chinese or Taiwanese system of medical education.

My rotations in my fourth year in clinical medicine were the highlight of my medical school experience. I was brought into contact with world famous clinicians and researchers. They ignited my interest in several areas in medicine. One was the endocrinologist at the Massachusetts General Hospital, Fuller Albright (1900-1969). Albright was at that time about fifty years old. He had what was called "post encephalitic Parkinsonism". This is a type of Parkinsonism, now extinct, seen after the widespread epidemic of encephalitis after World War I. The type of encephalitis involved is also thought to have disappeared; in any case its microbial cause remains unknown. Unlike regular Parkinsonism which afflicts older individuals, this type affected young adults. Albright was afflicted in his thirties. Its course was progressive, and it usually ended with severe disability. By the time I was with Albright, he was already severely disabled. He could only speak with effort and he could barely walk. But he had continued to see patients and work in science despite his disability. He gave me his secondhand car for the entire month I was with him. I had to take him to and from work every day. This way I got to know not only him, but also his wife and children. Another chore I had was to help him examine his patients. His disability was such that he could no longer do a physical examination. If I reported to him that I felt the spleen, he would palpate the patient's abdomen to confirm my finding.

In my spare time, I read all his papers. That in itself was an enjoyment. Albright's papers were famous. They were almost works of art. They were notable for clarity, originality, and wit. He invented many diagrams and cartoons to explain his thinking. He devised many methods of metabolic balance studies used to this day. Albright discovered many obscure endocrine diseases. One tongue-twister was named "pseudo-pseudo hypoparathyroidism". I was intrigued by how hyperparathyroidism resulted in the terrible spasmodic pain of obstructing kidney stones. Hyperparathyroidism is excessive production of the hormone of the parathyroid glands. This hormone is responsible for the maintenance of the precise concentration of calcium and phosphate in the blood stream. Excess amounts of parathyroid hormone produce a higher than normal concentration of calcium, such that the solubility constant of calcium and phosphates is exceeded and calcium phosphate is precipitated and stones are formed. Their passage through the kidney or ureter is blocked, and a spasm of pain is produced. This condition can be cured by surgically taking out one or more of the four parathyroids, which are located under the thyroid gland. Because of Albright's presence, parathyroidectomies was a common operation at the Mass General. A few years after I left Albright, he underwent a brain operation which was supposed to help him. It was Irving Cooper's "chemopallidectomy", in which small amounts of alcohol were injected in that part of the brainstem causing the rigidity and tremor of the disease. Cooper and Albright's own physician advised against the operation. But he insisted. The operation went well on the right side, but it was complicated by massive intracranial hemorrhage on the left side. He was left paralyzed, and he had to be sent to a nursing home, where he eventually died.

Albright taught me how to think about the mechanism of diseases.

The other professor who impressed me was the neurologist, Derek Denny-Brown. His service was on the ninth floor of the Boston City Hospital. He was a tall, lean, handsome but rather serious man, who spoke with the finest, clipped British accent. He seemed aristocratic to the core. During World War II, he was a general in the medical service of the British armed forces in the India-Burma theater. As a matter of fact, he was a New Zealander. I got the mistaken notion that all New Zealanders spoke with accent of the BBC type. Denny-Brown trained, directly or indirectly, almost all the foremost neurologists in the United States at the time. He was not only an outstanding clinician but he was a superb researcher, with a physiological bent. For example, he described the neurological complications of beri-beri, a vitamin deficiency. The entire service was permeated with his presence. I had the privilege of presenting a case to him. The resident in charge was more nervous than I was. At the end, Denny-Brown nodded his approval. Again, I spent most of my spare time during my month there reading the papers of his unit. At the end, I knew more neurology then than in the rest of my life.

I was fascinated by the precision of how signs and symptoms are produced in neurological diseases. The relationship between function and anatomic site of involvement has been beautifully worked out. Lesions, or anatomic abnormalities, may be traced to specific aggregations of nerve cells or nerve fibers located with pin point accuracy in the brain or the spinal cord. There is a condition we saw at the Boston City Hospital, called "Wernicke's encephalopathy", which may be found among large numbers of patients with advanced alcoholism, of whom we had many at the hospital. Patients are confused and they confabulate, that is they make false and illogical statements. But what is diagnostic of the disease is that pupils are fixed in position in midline. The eyes are unable to move to the lateral side. This is due to a weakness of the lateral rectus muscle, which enables the eye to move laterally. The nerve cells innervating this muscle are in the hypothalamus. This condition is caused by petechial hemorrhagic points, or minute bleeding points, in the midbrain and hypothalamus. The injury is due to a specific deficiency of a B vitamin called thiamine. What is remarkable is that this otherwise lethal disease is cured by a few shots of vitamin B.

By the end of my assistant residency, or the end of my clinical training, I had a stimulating experience learning and thinking about clinical medicine and disease mechanisms, but I had had no experience in medical research. There was in me an urge to get to the bottom of things, not unlike my urge when I wanted to seek the best political system when I first entered Harvard College. There were innumerable research groups at Harvard Medical School and its teaching hospitals. To look for a mentor in research, I first made an appointment with Dr. George Thorn, who was chair of medicine and was a famous endocrinologist. He was an authority on the newest and most popular group of hormones at the time, the corticosteroids. He was at the Peter Bent Brigham across town. Dr. Thorn had no vacancy. He recommended a number of his pupils in various parts of the country to whom he encouraged me to apply. I decided to stick to Harvard. I then interviewed with a number of other promising mentors. One was Dr. Emanuel Suter of the Department of Bacteriology. He was interested in the phagocytosis of tubercle bacillus. He found that exposure

The research staff of the Thorndike Memorial Laboratory, Boston City Hospital in 1957. First row, second from left, Edward H. Kass, Charles Davidson, William B. Castle, Maxwell Finland. Monto Ho is in the middle of the third row.

of macrophages, or large cells that ingest or phagocytize foreign substances including bacteria, to silicon dioxide severely reduced their ability to kill tubercle bacilli after they were ingested. This is an important observation because silicon dioxide is the most harmful part of coal dust, which causes the common coal miner's lung disease, pneumoconiosis. Coal miners were known to be susceptible to tuberculosis. Although this was an interesting topic in one of my areas of interest, infectious diseases, I decided to remain in a department of medicine so that I could get more clinical training and become a specialist. If I had studied with Dr. Suter, I would have become a basic scientist. Another investigator I interviewed with was Dr. Charles Davidson, a well known hepatologist, or specialist in liver diseases, at the Thorndike Memorial Laboratory at the Boston City Hospital where I had my residency[78]. But I found his research proposition was too general. The choice of a preceptor in research or a doctoral dissertation is one of the most important decisions in academic life. It is like a marriage. One can prepare for it, and one can get the advice of others. But whether it works out in fact often depends on interpersonal alchemy.

I finally decided on Edward H. Kass of the Thorndike Labs at the Boston City Hospital, who was a young assistant professor in the infectious diseases unit run by Maxwell Finland (1902–1987). Finland was one of the most important figures in infectious diseases in the United States. Like Davidson, he was one of the three main leading figures at the Thorndike Memorial Laboratory, and already a well known investigator. The third person was the chaired professor of medicine, a hematologist, William B. Castle. Ed Kass was a relatively new and upcoming investigator. Ed was the youngest of the "visits" on the second and fourth medical services, the two medical services run by Harvard Medical School. I was an intern and later a resident on the second medical service. His fame for brilliance preceded him. He had only one year of clinical training after medical school, when he became a member of medical faculty, and hence a "visit" or faculty supervisor of the clinical service. Their responsibility was to "visit" the ward in their charge three mornings a week when they were "on service". In city hospitals, the responsibility of "visits" was less formal than in private hospitals. All new patients were presented to the visit, and he was expected to visit and examine them. But the daily bread and butter decisions about taking care of the patients were left to the house staff, consisting of the chief resident, residents, interns and medical students. The lower echelons, that is the interns and medical students, were especially important, as they were expected to come up with all the decisions concerning diagnoses and therapy. The upper echelons, including the visits were more or less supervisory in nature. From Ed's rounding while I was on the medical service, he became known to the house staff for knowing everything about anything. His discussion of the cases during rounds was knowledgeable and erudite, but we were at times lost because they were not definite enough for us to follow his advice. We ascribed this to his relative inexperience as a clinician. That was in clinical medicine.

In research Ed was probably the most imaginative person I had ever met. His forte was command of the theoretical basis of medical phenomena. For instance, instead of talking about the signs and symptoms of an infectious disease and expecting us to recite them by memory, he would explain each manifestation on the basis of its physiological and pathological basis. He would explain many of the manifestations of gram negative bacterial diseases on the basis of endotoxin intoxication. A favorite example is the Waterhouse-Friederichson syndrome seen in bacteremia caused by the gram negative bacterium causing meningitis, the meningococcus. The patient presents with fever, shock, a hemorrhagic rash, adrenal insufficiency and death, unless quickly treated. Each manifestation can be explained by the effect of endotoxin on vascular perfusion, on the blood coagulation system and on the temperature regulation system. From this strong theoretical base, he would venture forth on treatment and therapy. The very early studies of treating pneumococcal pneumonia and bacteremia with corticosteroids in addition to antibiotics were an idea of Ed's, with Finland as coauthor. The use of corticosteroids in septic shock remains a debatable issue in therapeutic medicine to this very day.

Ed could go on for hours talking about his ideas and theories. At times he sounded a bit far fetched, but I later found out that Ed was extremely circumspect in what he wrote. He did not discuss all his ideas in his writings. The few ideas he chose to write about or set out to test were treated with integrity and responsibility in his papers.

While Ed was barely out of fellowship when I first went to him, he was constantly advancing rapidly. In 1988 I attended the 25th anniversary of the Channing Laboratory of Harvard Medical School, which Ed founded, and Ed's retirement as the first William Ellery Channing Professor of Medicine. In 1963, Finland and Kass were the two main figures in founding of the Infectious Disease Society of America, which has become the subspecialty society for infectious diseases in the United States.

By the 25th anniversary, Ed was a world famous man. He not only was a founder of the Infectious Diseases of America, he also founded the International Society of Infectious Diseases. He became a guru of the academic study of infectious diseases world wide. He had a whole corps of previous and present fellows. As his first fellow, I was on the program of the celebration and was one of the first to speak. I was asked for a vignette of Ed.

The image that came to my mind was from my fellowship with him. I was standing in his cluttered office on the 4th floor of the Thorndike Building, clutching papers with pieces of data in my hands, waiting to discuss them with him. His office looked like a mess; journals and books were piled high on the large desk and several surrounding tables. There were two telephones, both could be ringing at the same time; or he was answering one and the other one was ringing. I was waiting for him to finish one of his marathon telephone conversations, which often took a long time. Yet, Ed was always available, even if one had to wait, even in those early days. He was then already engaged in a remarkable array of activities, ranging from hospital rounds, not just in one but in several hospitals; to active involvement in medical school teaching, and he was the chief architect of the infectious disease core in Harvard's new third year integrated curriculum. That is, there were no departmental boundaries in the teaching. He was already in the process of planning for the new Channing building, a new institute of infectious diseases that he was to head, though at the time he was planning for it, no one knew who would run it. No doubt Ed was resourceful in seeking out opportunities, not only in scholarship and research, but also in personal advancement. He was also seeing patients with difficult urinary tract infections, which was his specialty in research. At the time he was doing a clinical trial of his new treatment for urinary tract infection, cranberry juice. He had reasoned that the short chain fatty acids in cranberry juice might be effective in killing bacteria in urine.

One of his famous papers[17] was the quantitative determination of numbers of bacteria in urine in diagnosing urinary tract infection. One of the difficulties in diagnostic medicine is the perennial presence of contaminating bacteria in uncatheterized samples of urine. Since catheterization was discouraged, being considered a mechanism by which bacteria can be introduced into and infect the bladder, for diagnostic purposes an overnight "clean catch" urine sample was obtained for analysis. However, no amount of cleansing could avoid bacterial contamination. One had to know what number of bacteria represented contamination and which represented infection. Using reasoning and simple tests in patients, he figured out these numbers. Since urine is a perfect medium for the growth of bacteria that produce urinary tract infection, and knowing the time it takes for a bacterium to divide (fifteen to twenty minutes), he reasoned that any infecting bacteria in the urine in the bladder would have been there for over five or six hours, as would be the case with overnight samples. One contaminating bacterium would generate about 100,000 bacteria per milliliter of urine. His conclusion was that contaminating bacteria would number less than 100,000 per milliliter of urine and infecting bacteria would be greater. As a matter of fact, he found by experiment that infecting bacteria were usually greater than a million, often ten million. Contamination after cleansing could easily keep the number at less than 100,000. These simple quantitative determinations have stood the test of time. They are the rationale for using bacterial quantitation in diagnosing urinary tract infections. These classical experiments have always been admired by me as indicative of the success of simple physiological, quantitative thinking. I always used them in my own teaching in explaining the dynamics of bacterial numbers in urinary tract infection. Yes, Ed always had many irons in the fire. If anyone could be in two places at one time, it was Ed. But one would not know it talking with him. He was never rushed, always took his time conversing with anyone, and listened carefully what one had to say. Despite long waits in his cluttered office before I could see him, I was never annoyed or disappointed. One learned so much overhearing his conversations.

The Thorndike Memorial Laboratory where Ed and I worked was established in 1923. It was one of the best known medical research laboratories in the United States. George Minot (1885–1950), one of its earlier directors, received in 1934 the Nobel Prize. He discovered an effective treatment for pernicious anemia, raw liver! Pernicious anemia is a chronic anemia prevalent among people of northern European origin. It was uniformly fatal until Minot discovered its treatment. By ingesting large amounts of distasteful raw liver, the patient's life was saved from inevitable death. Later it was found that the active ingredient in liver was trace amounts of vitamin B12, which was shown to be an essential vitamin for humans. Vitamin B12 is now purified, and instead of ingesting raw liver, patients with pernicious anemia can be cured by receiving periodic injections.

The Harvard medical services and the Thorndike Memorial Laboratory at Boston City Hospital were like a meteor in the American medical educational scene. To paraphrase Churchill, "never has so little produced so much", in terms of personnel development and research accomplishments in the United States. To the distress of us alumni, all Harvard units at the "City", the second and fourth medical services, the fifth surgical service, the neurology unit and the Thorndike Memorial laboratory were shut down in 1973, because Harvard acceded to the wish of Boston University to have sole use of the Boston City Hospital as its teaching hospital. The Channing Laboratories, founded and headed by Ed was also associated with the "City". It too left the "City" and is now associated with the Brigham Women's Hospital of Harvard Medical School. The Thorndike laboratory was at its prime when I joined it.

What attracted me to Ed was his interest in mechanisms and understanding in infectious diseases. He had a Ph.D. degree beside an M.D. from the University of California, which was rather unusual at the time. But this is one indication of his broad education in biology. He was proficient in microbiology, biochemistry and molecular biology. He used all his knowledge in thinking and conceptualizing about diseases. At the time, many in infectious diseases were engrossed with therapy, antibiotics. I was not attracted to research in antibiotics, because it seemed too mechanical to me. I would do research in antibiotic resistance for the first time after my retirement, but that is because antibiotic resistance is a major problem in health care in Taiwan. Its importance was social and my objective was service. Dr. Finland was the father of modern antibiotic therapy research and Ed's boss. I had to ask his permission to have Ed as my mentor rather than Dr. Finland, because I did not want to be stuck in antibiotic research. This was unconventional and I was the first to ask to be Ed's fellow. Dr. Finland accepted my request. It was only later that I appreciated Dr. Finland as a person. Although I was Ed's fellow, Dr. Finland regarded me as part of his group, and I was included in all the activities of his laboratory. Being a lifelong bachelor, his personal concern for all his fellows was legendary. In 1956, Carol and I adopted our first child, Bettie Pei-wen. Before Christmas vacation 1956, he walked one day into my laboratory with an armful of gifts. Without saying a word, he put all of them on my desk. The gifts were for each person in my family of three. Among them was a tiny sterling silver comb and brush set for Bettie Pei-wen, which she has kept and treasures till this day. I was flabbergasted by Dr. Finland's generosity and thoughtfulness.

Knowing my interest in mechanisms, for my research, Ed introduced to me a problem dealing with endotoxin. Endotoxin is the name of the internal toxin of many gram negative bacteria. These are the bacteria that cause diseases like typhoid fever, dysentery, Salmonellosis, cholera, bacterial meningitis and others. One reason for studying it is to understand the manifestations of diseases caused by these bacteria, such as fever, shock and death. Endotoxin is the key to unlocking many bodily reactions in response to infection.

At that time, a physiologist, R. Grant, who was a well known figure in the study of the physiology of fever produced by endotoxin, reported that after endotoxin was incubated with serum, its fever producing effect was enhanced. Ed wondered whether other effects of endotoxin, such as its death producing property, might also be enhanced by serum. At that time I considered it difficult to initiate a meaningful research project. There were so many imponderables and uncertainties. I felt like "a babe in the woods". Unlike other professors I interviewed, Ed asked a question which had a definite answer. The rationale of the problem he presented was straight forward. I could even think of ways to proceed and plan for the investigation.

The understanding and study of endotoxin required me to become familiar with many different areas in microbiology. Endotoxin is part of the bacterial cell wall. One had to understand bacterial structure and bacterial metabolism. To understand its toxicity, it was necessary to dig into the physiology of shock and fever as well as the basic tenets of immunology. To learn about these various facets of endotoxin was a wonderful way to review all these topics. I delved broadly and deeply into the literature of endotoxin and related subjects. Before planning for the experiments, Ed insisted that I map out all the procedures and steps beforehand, and in addition I had to predict and evaluate the possible outcomes. Our experimental model used the death of rats after intracardic inoculation of endotoxin as the end point. Before we started, I had to prepare large amounts of endotoxin, which I extracted from large cultures of a strain of avirulent *Salmonella typhi*. This required using basic bacteriologic techniques, such as growing bacteria, picking bacterial colonies, and harvesting bacteria. The endotoxin was then purified by the cold chloracetic acid precipitation of Boivin. This required the study of biochemistry and using chemical techniques.

For the animal experiments, I had a standing order of twenty to fifty rats from Charles River Company every two weeks. I went to work in the laboratory and animal room without any technical help. The animal room was on the top floor of the Thorndike building. After the animals were delivered in large flat cardboard containers, I had to dispense them in individual cages. They were fed and watered by one or two caretakers in the animal room. I never worked with animals before. First I had to learn how to handle them without being bitten. This was actually quite easy with rats because they are quite tame, and are not liable to bite. This is in contrast to mice, which wiggle around and bite at every opportunity. One can simply take the rat by its tail and drop it in a jar with cotton at the bottom containing ether. After the animal was asleep, I injected the endotoxin intracardiacally. This took some training. It was quite simple after one learned the trick. One needed to use a proper needle, a short one inch 22 gauge one. Then one had to place the injection midline in the thorax beneath the sternum toward the heart parallel with the long axis of the animal. Easy withdrawal of blood assured me that I was in the heart. The endotoxin solution was then injected.

After about two months, I had a meaningful result. This was that after endotoxin was incubated with serum, deaths in rats were not enhanced, but they were reduced.

The next thing I did was to investigate the component of serum that produced this result. Very likely a serum protein factor, presumably an enzyme, was responsible for reducing the toxicity of endotoxin. At that time, the famous Harvard biochemist, E. J. Cohen, had during World War II discovered a method to separate various fractions of proteins in serum by the cold ethanol precipitation method. Ed got in touch with his laboratory and obtained these fractions in powdered form. I incubated various fractions of Cohen with endotoxin and discovered that one fraction, the so called fraction II and III, had the ability to reduce endotoxin toxicity. We thought we had discovered an enzyme that could detoxify endotoxin. Its significance was that this enzyme might be a natural defense mechanism of the body. This small discovery, coming less than a year after I began laboratory work, gave me great satisfaction. I presented it at a meeting of the entire Thorndike laboratory. It produced in me the self confidence to undertake other investigative problems.

But before the work could be considered complete, I had to write it up and see that it was published as a paper. I quickly wrote up our experiments as a manuscript. Little did I realize that writing a scientifically meaningful paper was not an easy task. Ed looked at my draft and declared that it was unsatisfactory. He told me that the most important and most difficult part in writing a scientific paper was the introduction and the discussion. In the introduction, one had to describe why and how the project was undertaken and the scientific background for undertaking it. In the discussion, one had to discuss the meaning and significance of the results. In order to satisfactorily write these two sections, it was imperative first of all to have a command of the relevant literature. So I was sent to the library again, even though I thought I had a pretty good grasp of the literature. After one week, I had completed a lengthy redraft of the paper. He went over this with me. Besides discussing the general structure and outline of the paper, he went over the paper with me sentence by sentence, before he considered it satisfactory and submissible[18]. He taught me skills in the use of the English language. The redraft was reduced to a few succinct pages. In this way I was initiated by him in the art of the scientific writing. This experience was undoubtedly an important phase in my education. In retrospect, it was the continuation of an education that began at Brooklyn Tech, where I first learned expository writing; and at Harvard College, where I wrote a bachelor's dissertation in philosophy and government; and at Stanford, where I was taught expository writing by Professor Fairman of the Department of Political Science.

After a year in research, I already felt like a veteran. I felt I knew my way in the laboratory and I liked working there. I had tasted its fruits and its satisfaction. I respected and liked Ed as a teacher. I also became his friend. I liked research and decided I would proceed along this route.

My talk with Enders came around this point. I decided reluctantly to leave Ed Kass, after working with him for only one year. He was very disappointed in my decision, but he understood. Neither he nor Dr. Finland objected. Dr. Finland was a close friend of Enders, having taken care of his first wife, who died of an extremely unusual complication of influenza, myocarditis. I had already learned that Ed was a generous person, and my leaving did not affect our friendship, which persisted until his untimely death from cancer in 1990.

The significance of Enders' work for which he got the Nobel Prize in 1954 was that it made possible research work with viruses, which until then was hampered by lack of proper methodology. It facilitated the development of the polio vaccines, the most pressing medical problem in virology at the time. The seminal work was the description in a paper by Enders, Weller and Robbins in *Science* that they grew poliovirus in human skin and muscle cell cultures[19]. This work was a conceptual jump. Before their work it was assumed that poliovirus was tissue "tropic", that is, since poliomyelitis is a disease of nerves, it was assumed that poliovirus had to be grown in nerve tissue. Nerve tissue is very difficult to culture, because nerve cells do not multiply. The three types of poliovirus could only be grown in humans or in primates, except type 2, which can also be grown in mice. The forbidding prospect was that the only way to grow the virus so that vaccines could be made would be to use monkeys, an expensive and limited resource. The second important contribution, also described in this 1949 paper, is that the growth of the virus in cell culture could be visibly determined by visual observation, under the microscope. In other words, the virus replicating in the cells of the culture produced "cytopathology", or a pathological change in the cells. The pathology was due to the fact that the cells were killed by the replicating virus; they shriveled up. Fat, round cells eventually became little nubbins. This visible cytopathology, provided a convenient indicator useful for the "titration" or a quantitative measurement of virus. In terms of convenience for virus titration, the usefulness of a single tube culture was equivalent to a whole monkey.

Thus, this paper made possible the rapid development of the Salk and Sabin vaccines, both of which required a source for the three types of poliovirus and a method of titration of the virus. In the early 1950s, epidemic polio was almost a yearly scourge. It would occur during the summer months. Thousands of children would get paralyzed, hundreds were paralyzed so severely that they could not breathe effectively and had to be put in iron lungs. During the epidemics, swimming pools, camps and schools were closed. Parents were afraid to let children out of the house. Communities were panic-stricken. It was dreadful. There is no doubt the Salk vaccine trials stopped a polio outbreak in its tracks. In retrospect, it is the Sabin oral vaccine, which was released later, that kept polio contained world wide and eventually eradicated. The oral preparation was cheaper and more convenient to use, particularly in underdeveloped countries. But in 1954 the Nobel Committee gave Enders, Weller and Robbins the Nobel Prize, and not Salk or Sabin, whose work was considered developmental and based on Enders' work.

Enders' work also provided a strong impetus for the development of all other aspects of virology, since what was accomplished in the study of poliovirus could be extended to most of the other viruses. It made possible modern virology. This point of view is not understood or shared by the general public. The folk hero of the poliovirus vaccine work not only in Pittsburgh but throughout the country was Jonas E. Salk. One of Enders' stories about himself is his conversation with the taxi driver who frequently took him home from work.

"How is the progress of your research?" the driver asked. "Perhaps you will discover something wonderful, like Dr. Salk."

Amazingly enough, Enders is not a hero even in Boston. There is now a John F. Enders Building in the Harvard Medical School complex along Longwood Avenue. Enders had his office and laboratory in that building during the last few years of his life. I visited him there one day. I did not know my way to the building and asked someone in the street. He told me where it was. Offhand, I asked him if he knew who Enders was. I drew a blank face. Nor did he know that Enders himself was in the building.

One reason may be that Enders was not, like his mentor, Zinsser, a publicist. Enders told me that he did not think Zinsser was a great scientist, but he was a great publicist. Finland and Enders were both uninspired speakers.

By the time I arrived in Enders' lab, his primary interest had shifted to other viruses than poliovirus. Using the same cell culture techniques that he developed, he isolated measles virus in cell culture[20], making further work with this virus possible. He and two other fellows, Sam Katz and Don Medearis were engaged in the development of a live measles vaccine, which eventually was accomplished by the Enders' group. During the two years I was in the laboratory, there were altogether four to five fellows, like myself. According to today's scale, Enders' laboratory was a relatively small one, run with New England frugality. It was housed in an old building of the Children's Hospital. His laboratory was the Children's Hospital's "Research Division of Infectious Diseases". Most of his fellows were pediatricians, even though he himself was a non-physician basic scientist.

There were altogether three or four technicians, a secretary and a "kitchen", staffed by a unique Latvian group who took care of the washing and preparation of the laboratory glassware. This was before the days of plastic disposables. Among them was Yanis, an accomplished ex-painter and actor from his native country. The laboratory was more like a family. Friendship, good cheer and a relaxed ambience prevailed. There were common tasks where all the fellows and technicians were engaged. One was preparation of human amnion cell cultures, which had to be prepared every few weeks. Amnions were obtained from placentas from the neighboring Boston Lying In Hospital, also a part of the Harvard Medical School complex. Fellows did the stripping of the amniotic membranes. They were then stirred in a solution containing trypsin, which is a digestive enzyme that can detach the cells from the membrane. The cells were collected by centrifugation and placed in nutritious medium. Small amounts of the mixture were placed in small tubes and incubated. In a few days, the cells adhered to the glass surface, flattened out and grew in a monolayer. The medium was a rich mixture concocted in the Enders' lab. One mysterious ingredient, considered essential, was beef embryo extract. We had to prepare the extract from minced beef embryos obtained from the slaughter house.

Each Christmas, there was a joyous gala party given by the "Chief" and his wife, Carol, who was a previous technician. The party was at the Enders' home in Brookline, which was a quaint, delightful, comfortable house. My lasting impression was of its living room. It was a spacious, wood paneled room, with a balustrated open second floor lined with book shelves and books. It seemed so appropriate for Enders the man. We all knew that one feature of this party was that each member of the laboratory was expected to read a poem, written for that occasion. For the first time in my life, I wrote a poem. I wrote about Enders' daily visit during his rounds at the lab. The "Chief" told me afterwards that it was a very good poem. I took that as a compliment, coming from a scholar of English.

Right from day one in the laboratory, I became immersed in virology. This was a new subject for me, but it was congenial for me learning it from Enders. His virology and microbiology was oriented from the medical point of view. When he was preparing for his Ph.D., he took the first two years of the medical curriculum with medical students at Harvard Medical School.

Enders would start the morning at work, which did not start until the gentlemanly hour of nine thirty or ten, by rotating to the desk of each fellow.

He would sit down beside me, and ask, invariably, "What's new?" Even when there was nothing new, he would sit for a few minutes and chat.

Enders' conversation was thought provoking and inspiring, unlike his public speaking. At the time I was with him, I was reading "The Dialogues of A.N. Whitehead", as recorded by Lucien Price[21]. Enders' and Whitehead's conversation were similar. It is a mixture of conservatism and radicalism which makes for creativity. For example, when discussing literature, he would remark that a too detailed description of sex makes for boredom. At the time he was concerned about over population and the threat of atomic war. He was concerned about the thinking that wars would not touch American shores, that they would be elsewhere like during the two world wars. Americans may be too complacent.

Fred Robbins, a co-recipient of the Noble Prize with Enders, regarded him as a great scientist, but "we valued him more for his qualities as a man, and I, at least, considered one of his more unique and valuable qualities to be simple good sense"[22].

Francis Sargent (Sarge) Cheever, a Harvard alumnus and a true Boston "Brahmin", vice chancellor of health at the University of Pittsburgh, whom I replaced at the Graduate School of Public Health when he was promoted, talked with me often about Enders, as we both were Enders alumni and Enders fans. He always referred to Enders simply as "the great man".

He would chat for half an hour or longer, "if something new" launched him into a more prolonged discussion. These daily contacts were the essence of my training with Enders. They not only formed the basis of my new knowledge, they formed for me an unforgettable human bond.

Virology is a branch of microbiology. In the beginning, the only thing that was known to be different between viruses and bacteria was that viruses were "filterable" and smaller. But virology was behind bacteriology in sophistication, primarily because of lack of proper methodology. After the discovery by Enders and colleagues that viruses could be cultured in the laboratory in cell cultures, this animal virology had a new lease on life. When I went into virology, explosive advances were taking place. Another source of modern virology were studies of so-called "quantitative biology", which is an aspect of modern molecular biology. This new type of biology was founded by many people after World War II, almost all of whom were non-physician basic scientists. One of them was Max Delbrück (1906–1981), a German physicist turned biologist, who had emigrated to the United States. At Cold Spring Harbor in Long Island, New York, he and his colleagues initiated investigations in quantitative biology, concentrating on the study of bacterial viruses. These are tiny parasites of bacterial cells. One of his tenets was that a successful scientific investigation requires a quantitative method. Physics is the best example of this rule. He thought that the study of bacterial viruses could satisfy this requirement in biology. The reproduction of bacterial viruses in bacteria is extraordinarily fast, because the multiplication of bacteria and its viruses take place in hours instead of the days or weeks required for animal viruses and cells. After overnight culture, one sees plaques, or small, round clear areas of "lysed" or dissolved dead bacteria, on a "lawn" of susceptible bacteria killed by bacterial viruses. This simple, enumerative method of virus titration was revolutionary. For a generation after the early fifties, Cold Spring Harbor became a stronghold of investigation in bacterial virology. After a few years, Delbrück's expectations were realized. He received the Noble Prize in 1969. Bacterial virology contributed to important biological insights of the twentieth century, such as the demonstration that DNA is the genetic material. This insight was the foundation of Watson and Crick's Nobel Prize winning elucidation of the structure of the gene, which consists of DNA. After I joined Enders' laboratory, as was my custom, I studied the relevant literature including molecular virology. Most of the investigators in Enders' laboratory, including Enders himself, entered the field of virology through the study of diseases or from the medical point of view. Therefore, molecular biology was to them a relatively new discipline. I not only read about it, but in the summer of 1958 I registered for the course in bacterial virology at Cold Spring Harbor. So I became the exponent of this type of virology in Enders' laboratory. I was a pioneer in Enders' lab. Nowadays, because of Enders' cell culture techniques, animal virology has become as molecular as bacterial virology. The two different tracks in virology have fused, and the original distinctions have become blurred.

In his conversations with me, I expressed an interest in disease mechanisms, a continuation of my interest in endotoxin. Enders presented to me a basic problem in mechanisms that he had been thinking about, although no one in the laboratory was working on it. It was related to the meaning of the injury and death of cells that viruses produced, the so-called "cytopathic effect". This effect was thought to represent a mechanism by which viruses produced disease, by injuring or killing cells.

He gave me a strain of attenuated poliovirus called the RMC strain of type II poliovirus, which had been adapted to grow in a chick embryo, to work with. "Attenuation" means that the virus had been made non-virulent. The repeated passages in an unnatural host made it unable to paralyze mice. Type 2 polio virus was originally the only type that multiplied in mice, a non-primate animal. Along with attenuation, there was a reduced cytopathic effect in human amnion and kidney cell cultures observable in the laboratory. He thought that if one understood what reduced the cytopathic effect, one could get at a factor that resists disease in the animal host.

I began to study and think around this problem, and to design experiments. The seminal observation I made in the fall of 1957 was that RMC virus suspensions produced in human kidney or human amnion cells were different from virus suspensions produced in the chick embryo.

When an undiluted suspension of RMC virus made in amnion cell cultures was inoculated in a fresh human amnion cell culture, no cytopathic effect was seen. A cytopathic effect was seen only if the inoculum was diluted. This is called a "zone" effect. RMC virus made in the chick embryos did not have this "zone" effect when inoculated in amnion cells. I set about trying to understand this paradoxical situation.

After a few months of work, I discovered that this RMC virus produced in cell culture a substance, separate from the virus, which if layered on a fresh culture could reduce the cytopathic effect of poliovirus and other unrelated viruses. With reduction of the cytopathic effect, there was a diminution of viral replication. In other words, I found a new virus inhibitor. This was very exciting. I had a good response to his perennial question on his morning rounds, "What's new?"

Enders named the factor VIF or "virus inhibitory factor". Here was also the answer to the problem that Enders' had posed to me. We had found a factor that reduced the cytopathic effect, and decreased multiplication of viruses.

VIF was not part of the poliovirus, since it was not made inactive by antibody against type II poliovirus. It was not the size of a virus, since it could not be spun down in the ultra-centrifuge. It was not sedimented at 104,000 g. It was a protein, since it was destroyed by trypsin. To demonstrate its effect, VIF had to be added to cells before inoculation of the challenge virus. The implication was that an intracellular viral replicative event was inhibited. VIF inhibited many kinds of viruses, not just poliovirus.

At the time, we knew about the work of A. Isaacs in England on "interferon"[23], and realized that it and VIF had many similarities.

Alick Isaacs was a young British virologist who was interested in respiratory viruses. He was studying the phenomenon of viral intereference,

John Enders and Monto Ho, at Enders' home, Brookline, Massachusetts, 1959.

where one strain of influenza interfered with the multiplication of another strain. Traditionally, this was thought to be some kind of competition between virus particles. Isaacs however showed that the interfering virus produced in cells a substance that interfered or inhibited the second or challenge virus. He called this substance "interferon", or the agent that interfered.

Enders wrote Isaacs in England, describing what I had found, and asked him for a sample of interferon for comparison. He promptly responded. I tested his interferon and found that it had no antiviral effect in our human amnion cell culture system. Therefore we felt confident that VIF and interferon were different substances. We published our results in the *Proceedings of the U. S. Academy of Sciences*[24] and in *Virology*[25].

Later, additional work in Isaacs' laboratory showed that interferon was not one particular substance but a class of substances. Interferon from cells of different species of animals were biologically and biochemically different. This was what Isaacs called the "species specificity" of interferon. It explained why the interferon which he sent us, which came from chicken cells, was not effective in human cells. Isaacs foresaw and publicized the immense implications of interferon. He saw clearly that interferon might be a new class of antivirals.

Although I discovered an interferon, Isaacs had earlier discovered the class of interferons. After we realized the true situation, Enders announced at a meeting publicly that VIF was not a new factor but a type of interferon. This forthright and honest admission is in accord with highest standards of scientific ethics. This story was recounted in a paper I published in 1987[26], where I said that I accepted Enders' interpretation, "even if I felt a slight pang in my heart". However, the fascination with interferon would remain with me for twenty years. I left Enders' laboratory in Boston in 1959 for the University of Pittsburgh, with a NIH grant to study it.

My three years with Kass and Enders established my style and attitude in research. I was interested in mechanisms. I was interested in laboratory science.

Kass and Enders perceived my interest and directed me to scientific problems that were "basic" in nature. But "basic" is an ambiguous word in present day biology. It does not apply to me if it means only "molecular". The problems I was interested in were disease-oriented. It was a continuation of my interest in the fundamentals of broad scientific problems. It was the same attitude that propelled me to study philosophy and government at Harvard College in search of the answer to the question, what was the best type of government for China, and what was the basic problem in social science. It is also the attitude I took in dealing with the antibiotic resistance problem and Enterovirus 71 outbreak in Taiwan after I retired. These approaches basically represent the scientific approach to problems. It is a process of thinking and reflection. It is something that matured while I was in Kass' and Enders' laboratories and has remained with me. It exceeded the confines of strict science.

My interest in the laboratory was emulation of Enders' style of research. Throughout his life, his academic life remained in the laboratory. He did not give up the laboratory for conferences, travels and meetings. He may have given up pipetting (using the pipette, a basic laboratory instrument) himself in the laboratory, and left it to fellows and technicians, but most pictures of him showed him with a microscope. Indeed, the cytopathic effect or cell injuring effect by viruses was one of his basic observations. It was a visual observation made under the microscope, and he used it constantly. I cannot say that I was able to follow him all the way. I followed perhaps half of the way. Enders remained the pure scientist. I was interested in health and disease. The implications were that I became involved in practical problems. It was related to my desire to serve.

Enders the man had a great effect on me. Talking with him was akin to going to a retreat. All extraneous matters were shunted aside, and our conversation seemed to me to be the center of existence. His ambience was the essence of equanimity. He induced reflection and peace. It is interesting that we never talked about spiritual things. He did have a spiritual quality about him. His impact on me may be similar to my father's. Both were great influences. But the two were so different, and they came at different periods of my life. They influenced different aspects of my development.

chapter

| 8 |

RESEARCH ON INTERFERON

In a 1984 review, Mathilde Krim, a relative new comer in the interferon field, noted there was a "current windfall in interferon research"[27]. She was partly responsible for this windfall. An ex-actress, married to a millionaire Hollywood movie producer, Arthur Krim, Mathilde is a socialite active in the circles of wealth and power, in addition to being a knowledgeable scientist who worked at the Memorial Sloan Kettering Institute for Cancer Research in New York. In 1975 she organized an international meeting on interferon in New York, inviting known interferon scientists from the whole world and notables from the research establishment. The highlight social event of the meeting was a gala dinner at her sumptuous Park Avenue mansion. The glamour of the event overawed us.

There had been many international meetings on interferon throughout the world since its discovery, almost twenty years earlier. The participants were a sophisticated but critical lot, because of the many unconfirmed reports of clinical effectiveness of interferon.

What electrified the participants of this meeting was the report by Hans Strander of Sweden. Hans Strander is an oncologist who works at the Radium Institute of the Carolinska Hospital in Stockholdm. He was a good friend of Kari Cantell in neighboring Finland, from whom he obtained leucocyte interferon. In 1969, Strander began injecting 3 million units three times a week in patients with osteogenic sarcoma. He had been doing this for six years, and was obtaining unbelievable results. Osteogenic sarcoma is an extremely malignant cancer, usually tragically afflicting children. It had an unusually predictable fatal course. The usual two year survival was only 15%, but after three injections of 3.5 million units of interferon weekly for a year and half, Strander's patients had a two year survival of 60%[28]. Strander frankly admitted that he had not done a controlled trial. Still, the audience believed his results and were impressed. The ambience of the meeting became one of excitement. Most of the attending scientists were guardedly optimistic. But some believed that the miracle drug for cancer had arrived.

Mathilde maintained that what was needed was a major scientific effort to prove the effect of interferon on cancer. As a result of her meeting, many, including people with power and authority, agreed. That same year, she convinced the National Cancer Institute to allocate one million dollars for interferon research. This was the beginning of what she called the "windfall" of interferon research. She contributed to it by establishing an interferon laboratory at Sloan Kettering Institute. Her husband was member of the board of trustees. She recruited a number of young promising investigators, like Bud Colby and Bill Stewart. She included one of my own previous fellows, Kris (Y.H.) Tan, who later became the director of the Institute of Molecular Virology in Singapore. Kris had worked with us on the mechanism of production of interferon, and showed that its production could be enhanced by antimetabolites.

Right from the beginning, interferon research was full of hope. It was a new and hot area in medical research. With some ups and downs, it remained so from the late fifties through the eighties. A group of about twenty to fifty investigators throughout the world, half of whom were in the United States, mined the promise of interferon. They used the latest methods of medical sciences to open up and pursue novel perspectives in interferon research.

Hope was based on the concept of a new class of antivirals. There were and are few antivirals, in contrast to the vast numbers of antibacterials. Interferon was not only an antiviral, but it had properties that made it an unusually attractive one. That it is a natural cellular protein seemed to imply that it was non-toxic, like serum albumin; and initially it was thought to be non-antigenic. This meant that it would not produce hypersensitivity, and its effect would not be aborted by the production of antibodies. Unlike the few known antivirals in the sixties, like amandatine, which were unsatisfactory because they were highly specific and active only against a species of a virus or even a type within a virus species, interferon at its discovery was known to be active against a surprisingly large number of viruses.

Some cancers have been associated with viruses. Interferon was tried against cancers from the start. Reports of successes were in the literature, but none seemed to be as solid as Strander's.

You might think that if interferon was such a promising antiviral and anti-cancer drug, that the only important thing would be to do proper clinical trials. They are essential to show that it works. But clinical trials were not the only important thing about interferon. Interferon was not just another presumptive drug.

The fact that it occurs naturally sets it apart from other pharmaceuticals. There is the broad question of its meaning in the function of the body. Specific questions arise, such as where does it comes from, how does it work, and what is its purpose.

Interferon was the first thoroughly studied cytokine. These are natural proteins made within the body by cells in response to various stimuli. They are like hormones, which circulate in the body and have specific functions. In the course of fifty years, 37 additional cytokines have been discovered in addition to interferon, and the number is still counting. Many of these cytokines, including interferon, mediate immunological reactions. They have revolutionized the whole field of immunology. Cytokines are a group of proteins made to order by our genes. Another group that received attention earlier in genetics were enzymes. Modern biology attempts to understand the function of genes by understanding the functions of the proteins that they code for. Still, interferon is unique among cytokines because it is the only one with antiviral activities.

After almost fifty years of research on interferons, not all of the many initial avenues of research have borne fruit. Many original research results were preliminary and had to wait for later studies to make them more definitive. In order to explain the significance of my research, I will place them in the perspective of the present state of the art.

Having been initiated in Enders' laboratory in this area before I came to Pittsburgh, I applied for a NIH grant for research on interferon for my new job before I left Boston. It was funded when I arrived in Pittsburgh in 1959. This period of research lasted about twenty years.

My work would not been possible without the collaboration and help of many colleagues. Two had remained with me during the entire period. John Armstrong and Mary Kay Breinig were in the department at GSPH when I arrived in Pittsburgh. John is an English gentleman who was a graduate of Oxford University. He is broadly educated in all fields, but especially proficient in virology and biostatistics. He and I have been intellectual companions in research for forty years. We have discovered the gift of fruitful communication with each other. It became a form of cooperative thinking, which is not only pleasant, but at times inspirational and enlightening. To

Monto Ho in his virus laboratory, Graduate School of Public Health, University of Pittsburgh, 1983.

us, it was essential for productive research. Soon after I came to Pittsburgh, John completed a doctoral dissertation under my direction. He soon rose through the academic ranks and became a tenured full professor.

Mary Kay is a native of Pittsburgh, who was a technician when I arrived there. She had a bachelor's degree from Duquesne University. She soon transferred to work with me in 1960. We worked together on my earlier papers, many of which were a product of work by the two of us. I noticed her intelligence and care and precision in everything she did. I encouraged her to go on with her studies. She left her job and became a graduate student under my supervision in 1969. After obtaining her doctorate, she went to Philadelphia for post-doctoral work on genital herpes in 1975. She returned to Pittsburgh after two years to my group. After both of us left the interferon area, both of us worked on herpesvirus infections after organ transplantation.

Other associates in research were among the faculty, post-doctoral fellows and graduate students. Among faculty were Richard Michaels, Bosko Postic and George Pazin; among fellows were S. Suwansirikul,

Kris Tan, Y. Kono, N. Maehara, Xien Gui and T. Minagawa; among graduate students were Sabet Mahdy, Marion Harnois, Catherine DeAngelis, Stanley Singer, Charles Schleupner, David Jeng, Lucille Rasmussen, Tony Lubinieki and Carolyn Nash.

I divide what was significant in our research under the following topics: mechanism of action, inducers and hyporeactivity, mechanism of induction, and clinical studies.

ˎ Mechanism of Action

My first task in Pittsburgh was the study of the mechanism of action of interferon. How did interferon inhibit viruses? Where in the steps of viral replication did it act? Viruses cannot be understood apart from cells. Cells are factories that make viruses. The virus must first be adsorbed by and penetrate the cell. Within the cell, the nucleic acid is separated from the rest of the virus (uncoupling). The parasitized cell, using the viral nucleic acid as the blueprint, makes more viral nucleic acid and viral proteins. These subunits are assembled into hundreds of intact virus particles and released from the cell.

I hit upon the idea of using a phenomenon discovered by J. J. Holland and coworkers, a group of virologists at the University of Minnesota, to help solve the problem. They discovered that, while the intact poliovirus does not grow in rabbit cells, which are non-primate cells, if the ribonucleic acid of poliovirus is extracted from the virus and inoculated in a culture of rabbit cells, poliovirus is produced[29]. The reason is that intact virus must be adsorbed to the cell first, in order to replicate. Rabbit cells do not come from primates and they have no receptors for poliovirus; but if the naked RNA is inoculated in the cells, the cells which take up the RNA will make one round of virus replication. The virus made, being intact, cannot go on to infect other cells. There is only one round of replication.

I extracted naked RNA from poliovirus and inoculated it in chick embryo cell cultures, which like rabbit cells, are "non-permissive" for poliovirus. That is, they cannot make polioviruses. Before inoculating RNA, one set of cultures were treated with interferon and another was left untreated. Both sets were then inoculated with poliovirus RNA. The interferon treated cultures only produced 0.8 to 1.7% of the virus of the controls. This degree of inhibition was consistent with the amount of inhibition which we got with this interferon when we ordinarily inoculated intact virus. The conclusion was that interferon did not affect the adsorption, penetration or uncoupling of the virus after cell infection, but some intracellular synthetic process initiated by the nucleic acid was inhibited[30].

↳The abstract describing this experiment was selected for oral presentation at the annual meeting of the American Society for Clinical Investigation at Atlantic City in 1961, which was an honor. I was elected by that society in 1962, which is also an honor.

The present understanding of the mechanism of action of interferon is that inside cells interferon activates two enzymes which go through a number of steps, to inhibit virus replication by two different pathways. One pathway ends up with the destruction of viral RNA. The other pathway is inhibition of viral protein synthesis. Both pathways result in inhibition of viral replication.

The first pathway is initiated by an enzyme called 2'-5'-oligo-adenylate synthetase. It polymerizes ATP (adenosine triphosphate) which in turn activates an endoribonuclease that breaks down viral RNA.

The second pathway starts with a P1 kinase, an enzyme. Kinases are enzymes that add on a phosphate group to a molecule. P1 kinase phosphorylates and inactivates a protein synthesis initiation factor, called eIF2. Lack of this factor results in inhibition of the translation of viral protein synthesis, which is the process by which amino acids are hooked together to become a polypeptide or protein.

So our work pointed to the area in which interferon acted. It is now understood in more precise, molecular terms.

Another question I addressed concerning mechanism of action of interferon was the reason for the weakness of interferon's antiviral action. Interferon is an immensely powerful antiviral in terms of its minimal inhibitory dose. A unit of interferon, or the smallest amount that can inhibit viruses, is five one millioneths of a microgram. This is an incredibly small quantity. So interferon is an extremely sensitive viral inhibitor. This should translate into its being a powerful drug. And yet virus infections are difficult to cure even with very large doses interferon, or millions of units.

My paper in 1962 explains this dichotomy[31]. The dose response curve of interferon's effect, or the effect of dose of interferon on its antiviral effect, is steep in the beginning with small doses but it is flat at the top of the curve with larger doses. That is, with higher and higher doses, there is little increased viral inhibition. Another factor is that the effect of interferon is reversible with large doses of virus challenge. Interferon in small doses is effective, but even large doses can not stop virus multiplication. That is the basis of its weakness as an antiviral drug.

Inducers and Induction Therapy

In the first ten years after the discovery of interferon, it was thought it would be shown to be effective against at least a number of virus diseases. When not a single one showed up, one wondered what the problem was. A common problem of early clinical trials was that human interferon was not readily available, and larger doses could not be tested. Other avenues of utilizing the interferon system were explored. One was the use of inducers of interferon instead of interferon itself as therapy. This was called "induction therapy".

Wheelock tried this in 1964[32]. He injected different types of harmless viruses to induce interferon in a patient with leukemia. Interferon was produced but the method was faulted because antibodies against the virus are made by the patient after injection. They inactivate the virus. Thus, a virus could only be used once for induction.

It was discovered in the sixties that besides viruses, a large number of other microorganisms, such as rickettsia, mycoplasma, Chlamydia, and various protozoan could induce interferon, either in cell cultures or in the intact animal.

Gledhill made the observation that an injection of bacterial endotoxin in mice "spared" or reduced the morbidity of virus infection[33]. He suspected that interferon was induced, but he did not prove it. In view of my earlier interest in endotoxin and its manifold biological actions[18], I decided to test this hypothesis. I decided on an animal that was well known for its high sensitivity to endotoxin, and which we had already developed for interferon studies, the weanling rabbit.

One to seven hours after an intravenous injection of a minimum 2 µg of endotoxin, a serum factor was found in rabbit serum which had all the known properties of interferon including species specificity[35]. We also found that a large number of *Escherichia coli* injected intravenously induced interferon. The biological importance of this finding was it might explain natural immunity produced by gram negative bacteria against virus infections. Stinebring and Youngner[36] also found that endotoxin induced interferon in mice, around the same time as we did our work in rabbits.

In 1967, Field and his colleagues, in a group of scientists led by Maurice Hilleman of the Merck Company, reported that minute amounts of a mismatched ribonucleotide, polyinosinic polycytidylic copolymer, made large amounts of interferon if injected into rabbits, much more than when viruses were used as inducers[37]. Hilleman calculated that much more interferon could be induced than what was available to inject. Hilleman was a well known virologist, renowned for his effective work on vaccines in the practical areas of virology. He championed use of inducers of interferon. Research in "induction therapy" became a fad.

The problem with inducers was that they were almost all toxic. There was a period of about fifteen years of frenetic activity to search for the non-toxic inducer. No real successful one has been found, although there were many candidates. Most of them were large negatively charged molecules. In that respect, they were similar to endotoxin. The toxicity of these compounds was also similar to that of endotoxin. The severe reactions after injection were low blood pressure, even shock, associated with chills, fever and transient low white cell counts.

Hyporeactivity

A characteristic of the production of interferon in cell cultures and in the animal was that the amount of interferon induced soon reached a plateau and did not continue to increase with dose of inducer[31]. We then discovered the phenomenon of hyporeactivity. A single intravenous injection of endotoxin or a virus in a rabbit would render a second injection hyporeactive, or tolerant. That is the second injection would not result in production of interferon[38]. Later on other inducers, such as mismatched polyribonucleotides were also found to produce hyporeactivity[39,40]. There was cross tolerance in the sense that the hyporeactive state induced by one inducer was also active against another inducer[38]. Its duration varied with different systems. In rabbits, endotoxin produced a hyporeactive state that lasted as long as a week. In cell cultures, it lasted about a day.

The phenomenon of hyporeactivity has some practical importance. Cantell tells about the many difficulties he ran into trying to maximize the amount of interferon that can be induced in a given suspension of human white cells[41]. His method was to use a virus to induce interferon. He discovered early that a culture could only be induced once. After that they were refractory to reinduction so that a suspension could only be used once.

Obversely, hyporeactivity also restricts using interferon inducers as an antiviral agent, as repeated doses of inducers are not effective.

This interesting phenomenon is as yet unexplained biochemically. One does not know whether the hyporeactivity we discovered in animals and what many others discovered in cell cultures are basically the same phenomenon. We do postulate a protein that controls production by inhibiting interferon production in our study of the mechanism of induction (see below). This protein is also hypothetical. It might be the mediator of hyporeactivity.

Mechanism of Induction

It was already apparent in the sixties that interferon is a cellular protein and not a viral protein. According to modern concepts of protein synthesis, this meant that an interferon gene of the cell is derepressed or activated directly or indirectly, by an inducer of interferon. Indeed interferon genes have been proven and isolated. But how they are derepressed, or the precise mechanism of induction is not well understood. It is unclear whether all the inducers act the same way.

Since endotoxin and polyribonucleotide induced interferon was more rapidly formed than virus induced interferon, it was thought they induced "preformed" interferon. The inducer merely "released" it from the cells, and new protein synthesis was not involved. The hypothesis of "released" interferon became popular, especially when it was incorrectly assumed that this type of interferon production was not affected by inhibitors of protein synthesis[42,43]. Later, work by our group showed that this dogma was incorrect. We showed that interferon induced by a polyribonucleotide, the mismatched polyinosinic-polycytidilic acid, in cell cultures, a type previously thought to be "preformed", was inhibited by inhibitors of protein synthesis[44]. The same conclusion was applicable to endotoxin- induced interferon. This was demonstrated in tissue slices from rabbits that had received endotoxin intravenously, as endotoxin did not induce in cell cultures[45]. We developed the hypothesis that all types of inducers acted similarly, at least insofar as protein synthesis is concerned. They all induced by stimulating synthesis of a new protein, an interferon.

Following this discovery, we proceeded to develop a general theory of interferon formation. The outline of the theory is that after induction, two proteins are made; interferon, and a control protein that inhibits or stops the synthesis of interferon.

This theory was developed on the basis of our using two types of metabolic inhibitors to study protein synthesis. The first type was an inhibitor that prevented the first step of protein synthesis, the transcription of the interferon gene to a messenger RNA that would marshal together the amino acids so that they could be hooked together into a polypeptide (protein). Actinomycin D is such an inhibitor. The second type was a direct inhibitor of protein synthesis, that is, an inhibitor that prevented the translation of messenger RNA into proteins, cycloheximide and puromycin.

Paradoxically, if we added actinomycin or puromycin a few hours after exposure to the inducer so that transcription of the interferon gene or translation of interferon messenger RNA had already taken place, interferon production was enhanced. This was thought to be due to the inhibition of the synthesis of the second protein, the control protein.

If we used polyriboinosinic-polyribocytidilic acid as inducer of interfereon in rabbit cell cultures, the enhancement of interferon production after properly timed addition and removal of actinomycin or cycloheximide, reached over a thousand fold of untreated controls. The enhancement was even greater, if a combination of actinomycin and cyclohemide was used. The same phenomenon was observed with UV-irradiated Newcastle Disease virus, a viral inducer of interfereon, although the amount enhanced by metabolic inhibitors was less. We attributed the enhancement of interferon production to depression of the control or repressor protein[46].

The postulated control protein has also been used to explain the hyporeactive state described above[39,40]. So far neither the protein nor the gene responsible for it has been isolated. It is still a prevailing theory.

Besides formulating a theoretical theory of interferon induction, our findings were utilized to enhance interferon production for clinical purposes. At that time, the use of interferon for clinical purposes was growing apace, but the supply of interferon was very limited. Until the demand for interferon was solved by production in bacteria using genetic recombinant techniques[47], large quantities of interferon was only available by personal courtesy of Kari Cantell, who used the entire blood collection system of Finland to isolate leukocyte suspensions to manufacture inactivated virus induced interferon. For human use, interferon had to be produced in human cells, and the inducer could not be a synthetic polyribonucleotide or endotoxin, since these may have unknown toxicities. We devised a system using human cell cultures, using a combination of metabolic inhibitors and inactivated virus as an inducer of interferon[48]. K. Tan, J. A. Armstrong, Y. Ke and I obtained a U.S. patent for this system of production[49]. But it was never applied in industry, as the methods of manufacturing interferon by recombinant techniques soon predominated, as it has also supplanted Cantell's leucocyte interferon system.

Clinical Studies

Establishing proof that interferon was clinically effective was a tortuous affair. Almost as soon as interferon was discovered, and laboratory evidence of its efficacy became widely known, there were reports of its clinical effectiveness. In the 1960s, American travelers in Russia reported that interferon made in the chick embryo was available to the public for the treatment of upper respiratory infections. This violated the first principle of interferon action, that interferon is "species specific". Chick interferon would not be expected to work in humans. Later, they reported that human interferon could cure influenza[50].

One reason why early clinical reports were often erroneous or exaggerated was that the principles of rigorous clinical trials were not adhered to. Many illnesses get well without treatment. Unless one used untreated controls to compare with the treated group, one is often misled to an optimistic conclusion. Indeed principles of "randomized controlled trials" were foreign to most countries outside of the English speaking world until recently. It is no accident that many of the early reports of success were from countries on the European continent.

The second reason for early failures was that the amounts of interferon investigators used were generally too small.

Adequate clinical studies could not be done without adequate amounts of interferon. This problem was singularly met by Kari Cantell[41] of Finland from 1963 to 1981, when recombinant interferon was introduced and interferon supply became readily available. But during those 18 years, Cantell mobilized the entire blood transfusion service of Finland to supply many scientists with interferon. In the beginning he offered his interferon free, later his method of preparation was transferred to the Finnish Red Cross Transfusion Service who sold it to the American Cancer Society, NIH, Interferon Society of Houston, and the Karolinska Hospital in Stockholm for ethical clinical trials. Although he did not completely purify the interferon he produced, it was semi-purified and concentrated by a method that included the use of alcohol precipitation. It had a concentration of a million units per milliliter, which was high enough to allow his interferon to be injected intramuscularly in tolerable volumes for the patient. He reported for example that in 1974, they made fifty billion units of interferon, which required white cells from 70,000 blood donors, and from present knowledge, this corresponded to 250 mg pure interferon[41].

The magnitude of the amount needed to be therapeutic in humans was only clearly shown by Tom Merigan of Stanford University as late as 1973. He undertook a blind controlled trial at the Common Cold Unit in Salisbury, England. He used 14 million units of interferon per patient to relieve the common cold by rhinovirus injected intranasally[51]. This is the magnitude of the amount of interferon needed to treat disease in humans. The amounts used in earlier reports were usually in the range of thousands of units.

Tom is an immensely gifted investigator who also had his medical residency on the Harvard medical services of the Boston City Hospital. He was chief of infectious diseases at Stanford Medical School during his entire career. We have followed each other's careers closely. He once said that he and I were "birds of a feather". Tom pioneered in practically all aspects of interferon studies, but especially in significant clinical trials. He published a series of significant articles on the efficacy of interferon in herpes zoster in patients with malignant diseases of the blood. Although interferon is no longer used in chicken pox, his group made the significant discovery that interferon was effective in hepatitis B, which is still in use today.

I was not involved with clinical trials of interferon until the seventies. The interferon we used in our clinical studies was contributed by Kari Cantell, free of charge. Kari is a charming man of slight stature, who speaks with a melodious Finnish accent, very reminiscent of the Hungarian accent. It may be because the two languages are related. As can be seen from knowing him and from reading his book, he is a man of principle[41]. Despite his having possessed a potential gold mine because he was the sole source of interferon, he refused to profit personally from interferon. He also practiced openness and honesty in science.

By that time, I realized that interferon's effectiveness would not be easy to prove. One difficulty that concerned me was that according to its mode of action, a cell has to be pretreated with interferon before it can act in a virus infection. To realize interferon's full potential, I thought we needed to test it in a model of infection before the onset of the patient's illness. This was clearly a difficult requirement in human experimentation to meet.

We did develop a model along these lines. We tested the clinical efficacy of interferon in patients we knew would get labial herpes simplex, or cold sore of the lips[52,53]. Labial herpes is due to activation of herpes simplex virus that is in the trigeminal ganglion at the base of the brain. The activated virus is transported from the ganglion via fibers of the trigeminal nerve to its place of innervation on the lips and chin where it creates an area of inflammation that is the cold sore. Herpes simplex virus ordinarily remains quiet and latent in the ganglion until it is activated by poorly understood mechanisms, such as an upper respiratory infection or a cold.

At the time, the chair of neurosurgery at our teaching hospital, the Presbyterian University Hospital, Peter Janetta, discovered a new method to control the pain of trigeminal neuralgia. Peter was a charismatic neurosurgeon. He was a large, stocky man, characteristic of a surgeon; quick, decisive, cooperative and a man with relatively few words. I was surprised that he later took on the job of head of the health department of Pennsylvania.

His operation was to dissect away small arterioles which impinge on the trigeminal ganglion and separate them from the ganglion. A piece of foam plastic was interposed between the ganglion and the arterioles to keep them separate. Relief of pressure on the arterioles had the magic effect of relief from pain. An unfortunate complication of this operation was that the manipulation of the ganglion during the operation reactivated the herpes virus in the ganglion such that about half of the patients developed labial herpes within forty-eight hours of the operation. For us, the beauty of this reactivated virus disease was that it gave us a model of predictable viral infections. Ordinarily, the onset of a viral infection cannot be predicted. And once one gets a patient with a viral infection, it is usually too late for effective treatment. If one had a predictable viral infection, it was possible to test whether interferon given before the onset of symptons had a preventive or ameliorating effect on the lesions. At the time, Janetta's operation was becoming famous in the neurosurgical circles of the world and his department became a world center for this operation. With his help, we were able to collect a sizable number of patients so that a controlled clinical trial could be undertaken. We divided our patients into a treatment and a placebo group. For treatment, we administered interferon just before the operation and observed the development of labial lesions in both groups. We showed that administration of interferon could indeed prevent the activation of herpes simplex. However, if the administration was delayed until after the operation, interferon had no effect. This clearly showed a type of limited efficacy. These results were reported in 1979 in a series of reports[53,54].

Besides labial herpes we also studied the effect of topical and injected interferon on genital herpes in males and females[55,56]. There too we demonstrated efficacy but it was limited and clinically not very useful.

Another interesting observation we made was the effect of local interferon on disseminated warts, caused by papillomaviruses. Disseminated warts is a rather unusual disease, and it was not possible to collect enough patients to do a controlled study. We demonstrated the effect of local intralesional interferon in this single patient. As a control of our therapy, we inoculated lesions with saline in the same patient. It was remarkable to observe how intralesional interferon melted away the warts[57,58].

The first observation of the effect of interferon on papillomavirus was made by Hans Strander, when he was treating malignancies[41]. He observed incidentally that a patient with a wart on his skin disappeared after injection of interferon. This observation led to its use for juvenile laryngeal papilloma, which has to be extirpated in patients to prevent them from choking to death. Interferon was effective instead of surgery, but the administration had to be continued indefinitely. The disease was not cured.

Such observations led to the present licensure of topical interferon in condyloma acuminatum, which is a venereal wart-like disease of the genitals caused by papillomaviruses.

In the eighties, we undertook the treatment of hemorrhagic fever with renal syndrome in China. For this we used recombinant interferon alpha donated by Schering Company. This disease is caused by a member of the Hantavirus group. This group contains a virus known in the United States to produce disease called "Four Corners's Disease". This is a fatal pulmonary disease that occurred in the periphery of four western states, where they joined together. Hemorrhagic fever with renal syndrome is an endemic disease spread by field mice in various parts of rural China, especially Hubei in central China and in the northeast provinces. It breaks out during the early fall months, when the harvest is in and the field mice proliferate in the presence of vast amounts of stored grain. Patients are young adult males and females who are admitted with fever, and a hemorrhagic rash, which in severe cases proceed to bleeding and shock. It is associated with a disturbance of renal function, initially characterized by oliguria, or reduction of urine volume. Recovery is heralded by onset large volumes of urine or polyuria. The mortality varies but is about 15–20%. I had in 1981–1983 a physician post-doctoral fellow from Wuhan, Gui Xi-en, who was head of the infectious disease service at the Second Attached Hospital of the Hubei Medical University. He spent three fellowship years with me in Pittsburgh, and was an extremely competent fellow. He worked with me on some of our interferon projects. Before he left, he and I discussed and planned a clinical trial using recombinant interferon. Gui went home and took charge of the project from China's side. He negotiated with the central government, and the provincial government agencies, to facilitate the import of the drug and permission to do the trial. Most important of all, he recruited the Chinese patients who had to sign a consent form to enroll in the trial, which I originated in Pittsburgh after review by our review board for human experimentation. The whole idea of individual consent by signature was a foreign concept in China. He explained that it was in many ways more important to obtain the consent of the family and the "dan wei", or work organization for which the patient worked. Despite these cultural differences, he overcame all the red tape and difficulties, and recruited over a hundred patients who either received the drug or a placebo. I went to Wuhan in November, 1983 to observe his work in progress. We went to the neighboring villages to visit and interview patients that he had treated. This was my first contact with Chinese patients in China! It was also an experience for Gui, as it was one of the first clinical trials in that part of the country. The result of our trial was that there was a significantly better outcome in patients who received interferon, but it was not enough to make a significant clinical difference[59].

The hopes engendered by Strander's work at Mathilde Krim's conference in New York have only been partially fulfilled. Interferon is no longer used for osteogenic sarcoma. His results were never proven by controlled trials. A flurry of activity was generated in treating all kinds of cancers in various parts of the country. One of the most extensive was the work of Jacob Gutterman at the M.D. Anderson Hospital in Houston. One of his group, J. Quesada[60] conclusively showed that interferon arrested a rare type of leukemia, hairy cell leukemia. The disease is not cured by interferon, but can be arrested with periodic administration.

None of the herpesvirus infections, such as chicken pox, labial or genital herpes, is being treated with interferon, even though they have been shown to be effectively ameliorated by interferon. This is because there are now other more effective antivirals. Interestingly, Kaposi's sarcoma, a tumor that is caused by a herpesvirus (herpesvirus 8), has been shown to respond to interferon, by a remission in 20 to 50% of the cases[61].

All interferons used are now made by recombinant techniques. Cell culture-derived interferon is no longer used.

After almost fifty years of research of the therapeutic potential of interferon for infectious diseases and cancers, the present state of its usefulness could not been predicted. Interferons are now licensed for five diseases in the United States: Hepatitis B (interferon α2b), hepatitis C (interferon α2b), condyloma acuminatum (interferon α2b, αn3), chronic granulomatous disease (interferon γ1b), hairy cell leukemia (interferon α2a, α2b), and Kaposi's sarcoma (interferon α2a, α2b).

chapter
| 9 |

THE VIRAL FIFTH COLUMNISTS

David K., a 22-year-old young man had a heart transplantation in 1984 because of cardiomyopathy. In cardiomyopathy the heart muscle is almost dead and a new heart is needed for the patient to survive. David was transplanted at the Presbyterian University Hospital by Dr. Bartley Griffith. He was immunosuppressed by the new combination of cyclosporine and prednisone brought to Pittsburgh by Dr. Tom Starzl in 1981. The procedure seemed successful but six and half months later, he was found to have two golf ball sized lumps beneath his chin. A biopsy of the mass revealed abnormal looking lymphocytes, looking like cancer. The odd part was that the cells contained a protein of Epstein-Barr virus (EBV), called viral capsid antigen (VCA). His immunosuppressive drugs were reduced and the size of the tumors decreased. A brief course of acyclovir, an antiviral effective against EBV, appeared to halt viral shedding when we tested throat washings for EBV, but it had no effect on the tumor. When the tumor did not disappear, he received a course of chemotherapy like a conventional patient with lymphoma. He did not respond. His tumor became a large, necrotic posterior nasal lymphoma and he died eleven months after transplantation.

The transplanted patient sits on the horns of a dilemma. Unless the transplanted organ comes from an identical twin, it will be rejected by the recipient like a foreign body. The body's immune system gets rid of it like it gets rid of a germ or any foreign body. The only way to get around rejection is to give immunosuppressive drugs. Medical progress in the last fifty years has produced some remarkable immunosuppressive drugs so that it is possible now to transplant major organs. But they all have a major side effect: infections. Immunosuppressive drugs are like a coin with two sides, on one side is preservation of the graft, on the other is the danger of infection. The two go together. The immune system cannot distinguish between a germ and a foreign graft.

One or more infections, whether big or small, minor or lethal, whether by a bacterium, virus, fungus, or a parasite will occur after almost every transplantation. Many of these infections are manageable. They can be treated by our large armamentarium of drugs. But many are so severe that even with effective drugs, the patient may eventually succumb. Some viral infections are more serious as there are not many effective antiviral treatments.

David had an unusual type of virus infection. It was a lymphoma caused by a latent EBV infection. This is one of the so-called "post-transplant lymphoproliferative disorders (PTLD)".

I was in my office at Scaife Hall at the University of Pittsburgh School of Medicine when one day in January, 1981, Thomas Starzl and a retinue of three or four walked in and greeted me for the first time. He had just arrived in Pittsburgh from Denver, Colorado, to become Pitt's new Professor of Surgery. Tom was tanned, lean, athletic looking and handsome. He usually preferred informal clothing and sneakers, but that day he wore a tie and a lab coat.

> "Infectious diseases are the most serious medical problem in organ transplantation besides rejection. I need your help in dealing with these problems."

Tom's speech is remarkably low keyed given his strong personality. He is socially charming.

What he was asking for was help in managing infections in his patients after transplantation. Since they are at risk from so many infections, I agreed to give him all the help he needed. Following my conversation with Tom, and with the consent of the chairs of medicine and surgery, Gerald Levey and Hank Bahnson, I fostered the development of a new specialty, post-transplant infectious diseases, to serve the growing transplantation program. Two of our staff members in my Division of Infectious Diseases were exclusively assigned to the care of infections in transplantation patients; Simon Kusne was assigned to the liver patients, and Steve Dummer was assigned to the heart and heart-lung patients. Both were at first fellows in our Division, specializing in infectious diseases in order to be board certified. Later on they became the first staff members in our Division to specialize in post-transplant infections. Simon is a native of Israel. He remained at Presbyterian University Hospital until 2003. Steve was a reporter for Newsweek in Vienna, Austria when he decided to go into medicine. He started studying medicine at the University of Vienna but graduated from the University of Pittsburgh. He is now associate professor of medicine at Vanderbilt University School of Medicine. Both have made careers of post-transplantation infectious disease specialty.

Since we took care of all the infections after transplantation, we had a good idea of their nature and extent. Each type of transplantation concerns a different organ, and the nature of infections varied. In the case of liver transplantation, the operative procedure is within the abdominal cavity. The common post transplant infections were caused by bacteria and fungi infecting the abdomen, the peritoneal cavity and the abdominal wall. Candida is a yeast like fungus. Fungi are involved because the patients receive so many antibiotics that the yeasts, which are resistant to the usual antibiotics, are preferentially thriving. In cardiac transplantation, the organ to be transplanted is in the thorax. The major organ contiguous to the heart are the lungs. Infections of the lungs, or pneumonias are the most common infections after heart transplantation. They can be caused by cytomegalovirus (CMV), a fungus called *Pneumocystis carinii*, and various bacteria not ordinarily known to cause pneumonias, such as Klebsiella, Enterobacter and Staphylococcus. The organisms are common organisms carried by the patient himself, on the skin, respiratory tract or gastro-intestinal tract, even if they have strange sounding names. Some of them, especially the bacteria, are easily transmitted from patients to other patients in the hospital. The bacteria involved were frequently resistant to our common antibiotics.

When I became chief of the Division of Infectious Diseases at the School of Medicine in 1971, my attention was directed toward prevalent and important infectious diseases in the teaching hospitals that were viral in nature. Infections caused by CMV were becoming common with the onset of kidney transplantations. Tom knew of our work in 1975, when we showed that CMV could be transmitted to the transplant recipient by the kidney graft. He was interested in our work on the pathogenesis of CMV infections in humans and in mice. It was apparent that Tom was interested in our research, because it had broad biological significance for his specialty. At the time we were interested in developing an assay for "cytotoxic lymphocytes" directed against CMV infected cells in transplant patients. Cytotoxic lymphocytes primed to recognize CMV were then a difficult research problem, but now they are recognized as an important part of host defense against CMV. Our assay required a small piece of tissue from the patient on whom we were going to do the test after the transplantation. From the tissue we could grow fibroblast cultures which can then be infected with CMV. Infected cells would serve as targets for the cytotoxic lymphocytes which we could get from the blood of the patient. The piece of tissue could only be obtained from patients during an operation. He happily agreed to give us such tissues. I was gratified by his enthusiasm and was happy that a major surgeon was so eager to collaborate.

Our conversation ended on the happy note that we had found each other at Pitt. He had found a chief of infectious diseases who was interested in all infections after transplantation, and I had found a surgeon who was interested in research in infectious diseases.

Tom was born in 1926 in LeMars, Iowa. His father was of German, and his mother of Irish descent. He graduated from Northwestern University School of Medicine in 1952. He completed his surgical residency at the Johns Hopkins University School of Medicine. He then went to work at the University of Miami in Miami, Florida.

In his autobiography, he said, "If I had come to Miami in search of surgical competence, I achieved it automatically by caring for vast numbers of patients over the period of twenty-four months. In seeking a purpose beyond this, something at a subconscious level seemed to point to the liver."[62] He estimates that he performed over two thousand operations in Florida, reinforcing his reputation as a surgical virtuoso. His fascination with the liver, and liver transplantation would be life long.

He and Francis D. Moore of Harvard University School of Medicine were the first to successfully transplant livers in dogs in 1960[63]. In 1963, he transplanted the liver in a human after his group in Colorado had perfected the azathioprine and prednisone combination immunosuppression regime for transplantation of kidneys. But by 1967, using this immunosuppressive regime, he still could only record in his book a "Pyrrhic" victory of seven liver transplants but only three survivals. Many attempts were made and there were many failures.

By 1975, Tom had done enough liver transplants to be able to review ninety-three of his liver transplants with his favorite English pathologist, K.A. Porter. What was awesome about Tom was his perseverance in the face of repeated failures. For a long time, Tom was one of a few surgeons in the world doing this operation, but undoubtedly he had done more cases than any one else.

Now that he had a promising new immunosuppressive drug regime, cyclosporine and prednisone combinations, he decided to leave the University of Colorado in 1980 because they no longer wished to support his plans for an extended liver transplantation program. Within a matter of days after Tom decided not to transfer to UCLA after a visit there, the best man at his wedding in 1954 during his Hopkins days, the chair of surgery at Pitt, Henry T. Bahnson, invited him to come to Pittsburgh. Bahnson pushed his appointment through the university channels, not without opposition. It was with a certain amount of anxiety and expectation when we heard he might be coming to Pittsburgh. Tom was world famous and highly respected. But the liver transplantation program was controversial. Was such an operation justifiable or worthwhile, given its low success rate, and the vast amount of resources required for a single transplantation? Regardless, Tom's condition for coming was that this operation would be accepted at Pitt.

There followed after our conversation in January, 1981, the heady days when Tom brought the new immunosuppressive drug regimen from Denver, Colorado and started liver transplantation in Pittsburgh. He not only brought a promising new immunosppressive drug, cyclosporine, which was discovered by Jean Borel of Sandoz Corporation of Switzerland. He also brought his own contribution, which was the method of how it could be used safely and effectively in transplant patients. The British transplant surgeon Roy Calne had earlier tested cyclosporine and thought it was useless. He reported that it was toxic to the kidneys, did not prevent rejection of the graft and that patients developed lymphomas. Calne used cyclosporine as the sole immunosuppressive agent. The drug was about to be abandoned when Tom got a sample of the drug from Sandoz to try out and resurrected it. His secret was the combined use of cyclosporine and prednisone, just as he had earlier introduced the successful combined use of azathioprine (imuran) and prednisone. Prednisone is a corticosteroid and reduces the inflammatory response which is part of the organ rejection process. It acts differently on immune reactions than either azathioprine or cyclosporine. The combined use of these drugs with prednisone was synergistic. Before coming to Pittsburgh, he showed that the combination of cyclosporine and prednisone was better than the usual azathioprine and prednisone regime which was until then used throughout the world for maintenance of kidney grafts. But the real exciting implication was that the combination of cyclosporine and prednisone would make liver and heart transplantation feasible. For almost twenty years, while these two types of organ transplantations were technically possible, they were not generally performed because they were not consistently successful. Patients died because the grafts were eventually rejected or they died of an infection. These procedures were waiting for a better method of immunosuppression. Tom was waiting a better method of immunosuppression to restart his liver transplantation program. The combination of cyclosporine and prednisone appeared to be what he was looking for!

Tom Starzl took Pittsburgh by storm after his visit with me. Already by the end of 1981, the transplant program took a giant leap forward. The kidney transplant program, which had already been in existence under Tom Hakala, tripled in the number of transplants that year. Tom Starzl started liver transplantation. Others in the Department of Surgery, under the leadership of Henry T. Bahnson, the chair, and Bartley Griffith, introduced heart and heart-lung transplantation. From 1981 to 1985, there were a total of 1,467 transplantations at the Prebyterian-University and Children's Hospitals, of which 1,214 were in adults and 253 were in children. The total number rose from 144 in 1981 to 449 in 1985. There were in total 770 kidney, 477 liver and 220 heart and heart-lung transplantations[64].

The road of liver transplantation at Pitt was rocky and erratic at first. The first four persons transplanted all died. Another case required 200 units of blood, depleting the blood bank in the entire Pittsburgh region. Many operations lasted more than thirty-six hours. But gradually, with the new immunosuppressive regime and further technical improvements, immediate survival after surgery went from the long unacceptable 20% to an acceptable 80%. In 1983, a "Consensus Development Conference" sponsored by the Health Care Finance Administration (HCFA), which paid the bills of the "End-stage Renal Disease Program", the N.I.H. and other federal agencies sponsored a hearing of the benefits and liabilities of liver transplantation. The ruling was that liver transplantation was not experimental and was curable for certain diseases. This ruling eliminated some of the reluctance of third party payers to pay for liver transplantation. But how to pay for these transplantations remains a refractory problem[62].

The public gradually accepted liver transplantation. At Pitt, the initial objections to liver transplantation dissolved as the medical center became world famous. It became the largest center of organ transplantation in the world. Surgeons from all over the world came to Pittsburgh to learn liver transplantation from Tom Starzl. Patients from all over the world came to Pitt to be transplanted.

This however did not mean that organ transplantation had become a trouble-free surgical procedure. The complications of immunosuppression, especially infections, remained paramount. They were by no means eliminated by the cyclosporine-prednisone combination. The problem was in fact made worse. One problem was PTLD, and lymphomas. Calne discovered it to be a complication of cyclosporine immunosuppression. PTLD was relatively a new problem, because when only kidneys were transplanted, it was quite rare. But with the newer transplantations, especially heart and heart-lung transplantation, it became alarmingly frequent.

By 1984, Starzl and the Pittsburgh group, which including us, had collected and reported fourteen cases of PTLD from patients who were immunosuppressed with cyclosporine and prednisone after kidney, liver, heart and heart-lung transplantations[66].

PTLD comes in three forms. The first group consists of a picture like infectious mononucleosis with enlargement of lymph nodes, liver and spleen and inflammation consisting of mononuclear cells (mostly lymphocytes) that are benign and polymorphous. That is they were of many shapes, not monomorphous, uniform in size and shape. This group was not malignant.

Group two also consists of a lymphoproliferative process with polymorphous cells. But it can be more widespread than group one, infiltrating the liver, kidneys, lungs, intestinal wall, brain, etc. The patient may succumb.

Group three develop a tumor with monomorphous, monoclonal (single heritage) cells. The tumor very often involve the gastrointestinal tract. The cancer cells may also have chromosomal changes, like a translocation of chromosome 8 to chromosome 14 characteristic of Burkitt's lymphoma. Some consider chromosomal changes an essential additional requirement for calling the tumor malignant.

Among the 14 patients, PTLD occurred most frequently after heart-lung and heart transplantations (33% and 6%). It was less frequent after kidney and liver transplantations (2%). The patients developed their PTLD two to eight months after transplantation. The most common presentation was in the gastro-intestinal tract. Five had perforation of the small bowel, and another three had tumor nodules in the gastro-intestinal tract.

Our contribution was description of the Epstein-Barr virus infection in these patients, which we believed was essential for the development of PTLD.

What motivated Tom to publish this paper as soon as possible was that in this group of patients, ten of the fourteen patients became tumor free just by having their immunosuppressive drugs reduced or stopped.

He concluded that the "much publicized cyclosporine lymphomas are relatively innocuous if appropriately treated"[66].

I feel that this conclusion was a bit optimistic. He and I looked at the problem in different ways. Is the glass half full or half empty? It is true that reduction of immunosuppression can be curative, but it is less likely to be effective in liver, heart and heart-lung transplant patients because their grafts are essential for life. These grafts could not be sacrificed like kidney grafts, because patients without kidneys could survive by going on dialysis. Patients with frank lymphomas, that is when their tumors had the pathological characteristics of malignancy, which includes their "clonality" and chromosomal changes, reduction of immunosuppression was less likely to be effective. Our case David K., illustrate my point. The problem of PTLD has not been solved. It remains a major problem in transplantation today.

We continued to study PTLD and its relationship to EBV infection[64,67]. We believe it is the usual cause of PTLD. By cause, we mean it is essential, that is, EBV is a necessary condition. Without EBV, PTLD cannot occur. Two lines of evidence are cogent.

First, EBV genomes are always in PTLD's. To detect the presence of EBV DNA in tumors of our PTLD's, tumor tissues were examined for hybridization with probes containing fragments of EBV genome. Hybridization means a genetic blend or "complementation", in which the bases, or nucleotides of the probe match the bases of the virus in the tumor tissue. The presence of hybridization means that EBV was detected.

All eighteen tumors examined from twenty-five adults and children with PTLD's hybridized with a fragment of EBV genome by DNA-DNA hybridization except one. This one negative case may have been missed because of sample inadequacy . The presence of EBV DNA in tumors suggests that EBV is an essential factor in tumor formation, although how it works is still unclear.

Second, primary infection by EBV was more important than secondary infection in producing PTLD. Previous studies of PTLD had mentioned EBV infection[65], but they had not clarified the significance of the primary and secondary infections. Primary infections are infections of initial onset in a previously unexposed patient. Secondary infections are reactivations of old, latent infection in the patient. These two types can be distinguished by changes in antibody titer in the serum of the patient against a specific EBV antigen, the "viral capsid antigen" (VCA). In primary infections, there is an "antibody conversion", or change from absence of antibodies in a serum sample before the disease in question to a measurable titer after the illness has occurred. In secondary infections, there is a rise in antibody titers following the illness. In order to detect these antibody titer changes, it was necessary to have serum from transplant patients before and after transplantation. We devised a surveillance system to do this. We routinely collected serums from all transplant patients before transplantation and at monthly intervals for six months after transplantation.

We studied the fourteen adult patients and eleven children with PTLD. By serologic tests against different types of EBV antigens, we showed that all had active EBV infection. Six of the fourteen (43%) adults had primary infection, seven had secondary or reactivated and one had chronic infection. The proportion in the transplant population without PTLD who had primary EBV infection was 2.3%. The difference between 43% and 2.3% is highly significant, showing that primary EBV infections is a major risk factor for PTLD. This means that primary infection is more dangerous than secondary infection. This important point was dramatically brought out in children.

Because children are more often seronegative and at greater risk for primary EBV infection, their risk for PTLD is higher. Ten out of eleven PTLD's in children had primary EBV infection (91%). This almost suggested that for PTLD to occur in children, it was necessary for them to be primarily infected with EBV.

Out of 1,214 transplants in adults and 253 in children between 1981 and 1985, there were 0.8% of adults who had PTLD. Five times as many PTLD's occurred in children, that is, 4.0% of the transplanted children had PTLD ($P < 0.0005$). This showed that patients at risk for primary infection were at greater risk for developing PTLD. This greater risk was not anticipated before large numbers of children received transplantation in Pittsburgh.

Biology of Latent Herpesvirus Infections

In the last fifty years, research has revealed latent viruses can act like fifth columnists. Ordinarily, they behave like well mannered citizens, occasionally producing mild, non-lethal illnesses, but when defenses of the body are down, they can produce dangerous diseases.

Viruses that can become latent are what we mean by viral fifth columnists. Latency means that a virus becomes hidden or non-active. It is interesting that viruses that produce problems in transplant patients are ones that produce latent infections. Of the many groups of viruses, "herpesviruses" are most common and notable in being latent. (CMV), oral and genital herpes simplex (HSV type 1 and type 2), Epstein-Barr virus (EBV) are hepesviruses. These viruses consist of polyhedron particles with twenty faces measuring 0.00000047 inches or 12 Nm in diameter. Practically every mammal has its own group of herpesviruses, which share common properties but are different. There are eight types of human herpesviruses. They cause diverse diseases as chicken pox, oral and genital herpes, and infectious mononucleosis. These are ordinarily non-lethal infections in people with normal immunity. A herpesvirus also causes Kaposi's sarcoma. That is a fleshy tumor that can be seen nowadays in AIDS (acquired immune deficiency syndrome) patients, whose immunity is by definition deficient or compromised.

All herpesviruses have the capacity to remain latent and produce recurrent infection. Chicken pox can serve as an example. The initial or "primary" disease of chicken pox is an acute disease of children lasting one or two weeks. There is a widespread skin rash consisting of blisters or vesicles. It is common and highly contagious. Before the specific vaccine was available, practically all individuals got it once. Patients are prevented from getting chicken pox again because their bodies stimulated strong immunity. The important fact is that not all the chicken pox viruses in the body are killed by this specific immunity. Some remain latent in the dorsal ganglia of nerves in the spinal cord. They remain there for the life time of the patient asymptomatically and they usually cause no problem. However in some patients, when his/her immunity is knocked out or compromised by old age, lymphomas, anti-rejection drugs or anti-tumor chemotherapy, these latent viruses in the ganglia are "reactivated". More viruses are reproduced. They travel within the nerve cells they infect by fibers of these cells to their respective skin segment on the surface of the body and break out in a painful local skin rash or blisters. This localized chicken pox is called shingles or herpes zoster. Herpes zoster virus is the same virus that originally produced the chicken pox. In a way, the original chicken pox infection was never cured. All herpesvirus infections are the same; they are basically incurable.

Infection by viruses does not always result in disease. In fact most infections are asymptomatic, even in immunologically compromised patients. We know this because we can find antibodies to many viruses and other microorganisms in our blood stream which never caused any symptomatic disease. The production of a specific immune reaction usually means that most of the invading virus or microorganism has been inactivated or killed, and the patient is immune. All viruses stimulate what are called inactivating or "neutralizing" antibodies; these are antibodies that can specifically inactivate a virus. This is a powerful function of humoral immunity. However, even though all viruses stimulate such antibodies after infection, not all viruses are killed or eradicated after such antibodies are stimulated. Some virus infections may persist. Chronic and latent infections occur in the presence of acquired immunity.

Since latent infections are incurable, the presence of specific antibodies is also a sign that a person is latently infected. Infections can be divided into whether they occur in persons who are immune and seropositive or in persons who are non-immune and seronegative. When they occur in non-immune subjects, such infections are called "primary infections". They are called "secondary infections" in immune subjects. It should be apparent why primary infections are more severe and more symptomatic than secondary infections. The prior presence of immunity in secondary infection moderates its severity.

Epstein-Barr Virus

There is more latent EBV infection in the general population than CMV, most of which is asymptomatic. In our adult transplant population before transplantation, 98% were seropositive for EBV, that is they had been infected with the virus. The seropositive rate is a significantly lower in the children before their transplantation, 49%[64]. This is because some of them have not yet had a chance to be infected.

EBV is the cause of infectious mononucleosis in the normal immunocompetent population, a not uncommon disease of young adults in social economically well off communities in the western world. It has been called "kissing disease" since transmission results from intimate oral contact. Like CMV, EBV infects most people asymptomatically during young childhood. Only in communities that are relatively well off are there enough individuals who are still seronegative for EBV and are susceptible to infectious mononucleosis.

EBV has an unusual virological property. It infects mature non-dividing B lymphocytes in the blood stream and "immortalizes" them, that is, EBV makes the lymphocytes divide indefinitely. It is apparent that this property of EBV is related to cancer production. For cancer is uncontrolled division of cells. Cancer consists of "immortalized cells". Indeed, EBV is associated with two cancers in immunocompetent individuals, Burkitt's lymphoma and nasopharyngeal carcinoma. They are cancers found in equatorial Africa and in southeast Asia.

Where did the virus in primary EBV infection after transplantation come from in our patients with PTLD that we described in 1984,1985 and 1986? In view of our experience with CMV and herpes simplex virus, we naturally suspected the transplanted organ graft as the source. But we had no proof.

In 1991, we had the opportunity to directly demonstrate the transmission of the virus from the donor to the recipient by molecular methods. This was made possible because of the arrival of a young molecular biologist in my group at the Graduate School of Public Health, Jennifer McKnight. She had a Ph.D. in microbiology. Jennifer was a beautiful, talented and artistic woman. She was meticulous and demanding in her work. Like a meteor, she came and vanished from our lives after a mere three years with us. She suddenly quit the academic scene. Before she left, she trained a gifted but initially undisciplined Chinese graduate student, Hui Cen. Together, they showed how EBV was transmitted[68]. It would not have been possible without Andrea Zieve, a transplantation immunologist who happened to have a sample of the spleen of the donor, which was essential for the success of this study. This type of tissue would not ordinarily be available. It happened to be possible in Pittsburgh where all aspects of research relating to transplantation was then thriving.

The blood of Patient T was seronegative for EBV before a heart-lung transplantation. His donor was seropositive. The donor EBV was available from a sample of spleen tissue of the donor that was saved by Andrea. The EBV of the recipient was obtained from a lymphoblastoid cell line that was developed from his blood cells. The viral genomes from the spleen and the cell line were expanded by polymerase chain reaction (PCR). PCR is a remarkable chemical reaction which simulates the reproductive process of nature. That is, given the chemical constituents of DNA in a test tube, and the beginning and end pieces of a stretch of DNA, thousands of copies of the entire stretch of DNA is made. The stretches of DNA made can then be characterized by probing for hybridization or identity with "IR3", a highly variable portion of the EBV genome which is unique for each strain of EBV. The virus from the donor and the recipient was analyzed by "restriction endonuclease fragment length polymorphism" (RFLP). This is a test that finger prints genominal DNA now commonly used in criminal investigations to identify individuals. DNA is broken into fragments by the action of specific enzymes, called DNA restriction endonucleases. The fragments are separated in a gel by electrophoresis according to molecular weights. The pattern of sedimentation of the fragments is unique for each individual genomic DNA.

The viral DNA from the donor and from the recipient were found to be identical. Therefore we believe that the recipient virus came from the donor[68].

Having the virus from the donor enabled us to prove directly transmission of primary infection of Patient T by the graft.

The Story of Cytomegalovirus (CMV)

Of all microbes and herpesviruses, CMV produces the largest number of illnesses in transplant patients. Like EBV, CMV is a herpesvirus.

Until Tom Weller succeeded in culturing cytomegalovirus (CMV) in 1957[69], CMV was not known as a herpesvirus. It was then known as an esoteric virus that caused an extremely rare, fatal disease of new born infants, "cytomegalic inclusion disease". Its name came from the fact that tissues from patients with this disease showed, "cytomegaly", or large cells. They have characteristic structures within their nuclei, called giant intranuclear inclusions, which are loaded with CMV virus particles. Up to then, the layman never heard of the virus, and even doctors knew little about it.

Epidemiological studies were made possible after Weller's isolation. Antigen could be made from the virus isolated, and serological tests could be undertaken to measure the CMV antibody titers in population groups. These titers reflected past infections with CMV. Surveys revealed that past latent CMV infection was very common among humans. In some population groups, it was over 90%. Even in communities with low infection frequencies, over half of the adults were infected. Humans were usually infected, without symptoms, during infancy or at a very young age. Nowadays, we believe that infection is transmitted by human milk from the mother or from milk pools. Another source of CMV are "fomites", clothing, bedding, toys, contaminated by CMV containing urine of toddlers in day care centers, or where large numbers of small children aggregate.

Even though CMV was a relatively unknown virus, we now realize most of us have been infected and remain latently infected by it.

By the 1980s and 1990s CMV had become commonplace as an agent causing disease. CMV retinitis became the most common and devastating cause of blindness in AIDS patients, and occasionally in transplant recipients. CMV became a ubiquitous problem of organ and tissue transplantation. Among patients who received kidney, liver, heart, heart-lung and bone marrow transplantation, 33% to 70% became infected with CMV, not necessarily all symptomatic. The incidence of CMV pneumonia, the severest form of CMV disease, ranged from 32% and 16.7% in heart-lung and marrow recipients to 2% in kidney recipients. For a brief period in the beginning of the transplantation era, patients with kidney transplants were isolated, and environmental precautions were taken to prevent transmission of outside pathogens. It soon became apparent that such precautions were ineffective. One concluded that CMV did not come from the patient's human or material environment.

Since many of us are already infected with CMV latent in us, the source of CMV that causes illnesses might be the patient himself. This is indeed the case in AIDS retinitis. The CMV causing the retinitis is a latent virus that has become active. One had to determine how important reactivation infection was in the case of CMV infection in transplant recipients.

One also had to determine how important primary infection is, and one had to determine the source of CMV in the patients who had primary infection. That was a great mystery.

It was at this point of thinking in 1971 that I began our studies on CMV. There was only kidney transplantation at the time. The following patient illustrates what we saw in patients.

Alice N.'s whole life had been preoccupied with her kidney disease. As a young girl, she had urinary tract infections. At that age, these infections involve the kidney and growth may be stunted. Failure of her kidneys developed. Now a young woman of 20 years old, she had to go to the hospital once a week to be put on dialysis in order to get rid of excess waste products in her blood. This procedure kept her kidneys from continuing failure and death but it could not be continued indefinitely. She had been put on the waiting list for a kidney transplant. Finally, she was transplanted with a kidney from a newly dead donor. The procedure went well. When she woke up, she had a new kidney. She no longer needed to go to dialysis and her life was full of hope! The only hitch was she had to take immunosuppressive drugs, azathioprine and prednisone indefinitely so her body would not reject the foreign organ. She was discharged from the hospital. Her recovery at home seemed to be on course when three months after transplantation, she developed fever, a hacking cough and shortness of breath. Her chest X ray showed diffuse spots in the lower lobes of the lung, signifying pneumonia. The oxygen level was low in her arterial blood because the lungs could not supply enough oxygen. She needed respiratory assistance for delivery of oxygen. This could only be done in the intensive care unit where she was transferred. She was placed on the critical list. To help her immune system, the doses of her immunosuppressive drugs were reduced. Her blood and urine cultures became positive for CMV. Her antibody test for CMV which was negative before her operation was now positive. This meant she had a primary or first incidence infection with CMV. In order to determine the cause of her pneumonia so that proper therapy could be given, a lung biopsy was performed by inserting a hollow needle through the chest wall into the lung. A sliver of tissue was obtained. Under microscopic examination, the biopsy showed large cells with large intranuclear inclusions, typical of cells infected with CMV. A homogenate of the biopsy material was inoculated in human fibroblast cultures. The cells of these cultures developed morphological changes indicative of CMV. The pneumonia was caused by CMV. Unfortunately, there was no antiviral drug against CMV in 1971. After a stormy course of two weeks in the ICU, she eventually recovered, and went home.

By happenstance, the NIH was interested in determining the risk of CMV illnesses after blood transfusions which were thought to be able to transmit CMV. Our research laboratory responded to the NIH request for applications (RFA) for contracted research. I believed that this type of research could help us in care for patients with CMV illnesses.

We at the University of Pittsburgh were exceptionally qualified because we had many types of patients with CMV infection. I organized a strong research group to address this clinical question. Beside myself, John Armstrong in our faculty of the Graduate School of Public Health was a trained virologist. In addition, we recruited Donald N. Medearis, who was dean of the medical school and a professor of pediatrics. He was my contemporary while we were both in training in the laboratory of John Enders at Harvard Medical School. He had done some pioneer work on CMV. He brought with him to Pittsburgh and into our research group, a veteran laboratory technician, Leona Youngblood, who had worked with him in Boston and was versed in all laboratory techniques in CMV. John N. Dowling, a new faculty member in our Division of Infectious Diseases, and a fellow in infectious diseases, Sakdidej Suwansirikul, made up the six members of our team. "Suwan" was from Thailand. During his two year fellowship with us, he was quiet but extremely able.

We also had exceptional clinical materials to work with. Our Children's Hospital had a thriving program in open-heart surgery, mainly dealing with the correction of congenital anomalies. Open heart surgery requires extracorporeal by-pass perfusion of blood. This requires a significant amount of transfused blood. We included two other clinical populations with CMV infections in the University of Pittsburgh Medical Center in our studies; renal transplant patients, and patients placed on cytotoxic immunosuppressants in the rheumatology clinic. We responded to the RFA and were funded for three years. This was the beginning of a subsequent continuous effort for more than twenty years, during which we made numerous contributions. I wrote a book on the biology and diseases of CMV, which went through two editions[70].

I discussed the problem of CMV disease in renal transplant recipients with Dr. John E. Craighead at an academic meeting in 1971. He was the chief pathologist at Dartmouth College School of Medicine in Hanover, New Hampshire. He had worked with the pioneer surgeons of kidney transplantation at the Peter Bent Brigham Hospital of Harvard Medical School, including Dr. John E. Murray, who later received the Nobel Prize in 1990. In 1965, Dr. Craighead and his colleague had described the first case of generalized cytomegalic inclusion disease in a renal transplant recipient[71]. They were one of the few early workers in transplant medicine who had written about CMV. They also had mentioned but did not prove in their paper that the transplanted kidney might be the source of CMV[71]. John thought that infection with CMV almost always occurred after renal transplantation. It was inevitable but he thought it was rarely serious. He did not encourage me to go into the area. John had grossly underestimated the importance of CMV. At that time, the renal transplant recipients who got only azathioprine and prednisone without additional immunosuppresants such as the later discovered anti-lymphocyte serum did not usually get severe CMV disease. Our patient Alice described above was about the worst type of illness they got. He did not then realize that later more potent immnunosuppressants, such as the anti-lymphocyte serums, and other types of organ transplants, such as bone marrow and heart transplantation would bring on the real devastation of CMV.

I was not discouraged by John. I thought it was important to distinguish the two types of CMV infection after transplantation, the primary or initial onset type, and the latent reactivation type, and to characterize the clinical illnesses associated with each. This had not been done. I was also interested in proving that primary infections came from the donor kidney graft. Since CMV could not be demonstrated in the donated kidney, I thought of a way to get around that difficulty in another way.

Before our studies, we too tried to culture the virus from samples from the kidney graft and failed. Even today, almost thirty years later, demonstration of latent virus in normal organs is still not possible even with modern molecular methods such as PCR (polymerase chain reaction), which makes possible the detection of minute amounts of DNA or RNA. We took another approach. We reasoned as follows: if a patient is infected with cytomegalovirus for the first time, he/she will seroconvert, or the specific antibody titer in blood will change from negative to positive. If seocon-version takes place after transplantation, primary or *de novo* infection has taken place. We also postulated that if the donor of a kidney is seropositive, it would mean that his kidney is latently infected with CMV. If a seroneg-ative recipient gets a kidney from a seropositive donor and seroconverts, this would be presumptive evidence that infection had been transmitted by the transplanted kidney. To test this hypothesis, we decided to prospec-tively test a group of seronegative recipients who were transplanted with kidneys from seropositive donors.

We followed in 1972–1974 fifty transplant recipients prospectively who had survived at least four weeks after transplant operation. A serum and urine specimen were obtained from the donors. Blood and urine specimens were obtained from the recipient before the operation and at monthly inter-vals for six months after the transplant. All patients were followed for six months after transplantation.

Ten out of twelve, or 80% of seronegative recipients who got kidneys from seroposistive donors became infected, but only three out of ten or 30% of seronegative recipients who got kidneys from seronegative donors became infected. The difference was statistically significant. This suggested that kidneys from seropositive donors were latently infected with CMV. The virus was transmitted to the seronegative recipients causing in them primary infections, and the virus became active after transplantation.

Out of the ten primary infections, seven of them or 70% had CMV disease after being infected. Secondary or reactivation infection, or infections in initially seropositive recipients, occurred in sixteen out of twenty patients (80%), but only two were ill (10%). This showed that primary infection was distinguishable from reactivation infections, and primary infections produced more symptomatic disease.

This work was published in 1975 in the *New England Journal of Medicine*[72].

In a later study of fouty-seven renal transplant patients, we reported the differences in clinical manifestation between primary and reactiva-tion infections. There were eighteen cases of primary infection and ten cases of reactivation infection, based on whether the patient was seroneg-ative or seropositive before transplantation. Thirteen patients with pri-mary infection (72%) and only one with secondary infection (10%) had symptoms of illness. Illness was defined as two or more manifestations of the following abnormalities: fever, leucopenia (abnormally low white cell count), atypical lymphocytosis hepatosplenomegaly (enlargement of the liver and spleen), arthralgia (pain in joints), and pneumonia. One patient with primary infection died with CMV disseminated disease[73]. In renal transplantation where CMV disease is milder than in other types of transplantation, one can say primary infections are clinically the only important ones unless anti-lymphocyte serum is also used for immunosuppression.

The main points of our work, transmission of the virus by the kidney graft and the importance of identifying primary infections because of their increased clinical severity, have been amply confirmed by other workers. It is now accepted practice to measure the risk transmission of CMV from the graft and of primary CMV infection by determining the CMV serology of the donors and recipients. One now knows seronegative recipients receiving organs from seropositive donors pose the highest risk.

As we examined data from other types of organ transplantation besides kidney, we assumed that these organs were probably also latently infected with CMV. The rates of primary infection after kidney, liver, heart and marrow transplants were 52%, 46%, 64% and 43%. These rates are high enough to assume that the transplanted organ transmitted latent CMV to the recipient, where it reactivated. In all patients, the morbidity, or degree of illness, was higher in primary than in secondary infections. In heart transplant patients, where CMV disease is most severe, 90% of the patients with primary infections were ill, while only 29% of those with secondary infections were ill.

Our experience at the University of Pittsburgh showed that CMV disease was more serious after liver, heart, heart-lung and intestine transplants than after kidney transplantation. The frequency of CMV pneumonia was 8% after heart transplantation, 32% after heart and heart-lung transplantation, and 2% after renal transplantation. The reason for this is not differences in immunosuppression but the difference was probably due to the nature of transplantation itself[70]. The importance of the type of transplantation is best shown in bone marrow transplantation. There a unique graft versus host reaction might take place which is associated with CMV pneumonia of an unusually high 90% mortality and a frequency of 17%[74].

Under ideal circumstances, in order to prevent primary infection, one would only transplant organs from seronegative donors to a seronegative recipient. This however is not possible because of the shortage of donors.

Still the prior identification of seronegative recipients who receive organs from seroposistive donors has become essential for identification of the high risk group for severe CMV infection. This group is often selected for prophylactic or pre-emptive antiviral treatment with ganciclovir. This is an antiviral drug effective against CMV that became available after 1985.

Herpes Simplex Virus

This herpesvirus is probably the best known one among laymen. Herpes simplex virus (HSV) comes in two types. Type 1 is the cause of the common cold sore, a recurrent eruption of blisters on the lips and chin which lasts a few days to about a week. Type 2 causes a similar eruption on the genitals. Viruses that are hidden and latent reactivate as in the case of chicken pox virus discussed in the beginning of this chapter. The latent viruses of type 1 are in the neurons of the trigeminal ganglion, or the ganglion of the fifth cranial nerve. When they reactivate, the virus becomes active, reproduce and travel by way of the fibers of the nerve to emerge on the skin of lips and chin that are innervated by the fifth nerve. Similarly, the virus of type 2 in the lumbar sacral ganglia of the spinal cord travels to the area of the skin of the genitalia that is innervated by these ganglia. On the skin, the viruses stimulate a blister forming inflammation. Recurrent episodes of eruption can take place in an individual who is immune. The primary or initial infection by these viruses may be asymptomatic or symptomatic. When symptomatic, the disease picture is more severe; it may involve a wider area of skin eruption, or dissemination throughout organs in the body. It may also be accompanied by fever.

Unlike CMV, which appear to be latent in many organs and tissues, latent HSV is restricted to certain neural ganglia which are not transmitted during transplantation of organs. Therefore, primary HSV infection after transplantation is not common. However, from our surveillance of herpesviruses we came upon the two cases described below.

Bruce M was a 47-year-old man with polycystic kidneys. This is a congenital deformity inherited in an autosomal dominant manner. That is, if either parent has the disease, the responsible gene is transmitted and manifested in half of his offspring. The victims are asymptomatic until midlife. Then the cysts on the kidney enlarge and fuse together. The kidneys become so large that they can be easily felt by palpating through the abdominal wall. They are easily traumatized so that there is bleeding within the kidney.

The cysts obstruct the function of the kidneys and renal failure ensues. Like Alice N., Bruce received a renal allograft and his polycystic kidneys were removed. Twenty days after transplantation, he developed blisters or vesicles on the skin of his hands, feet, and trunk. Endoscopic examination showed multiple ulcers, 3–4 mm in diameter, in the antrum, or the top portion of his stomach. He was treated with acyclovir, an antiviral drug effective against herpes simplex, for ten days, and his symptoms resolved. Cultures of his urine and vesicles all yielded herpes simplex virus (HVS) type 2.

Carol L. was a 29-year-old woman with chronic renal failure similar to Alice N. She received a kidney transplant from the same donor as Bruce M. Urine specimens obtained three days after transplant yielded HSV 2. Eighteen days after transplantation, the patient developed fever, and a disseminated vesicular rash involving the hands, chest, shoulders and abdomen. A biopsy from a vesicle and a urine and throat specimen yielded HSV 2. Liver function test were slightly abnormal, which is a common finding in disseminated HSV disease. She was also treated with acyclovir for ten days and recovered.

Both patients were seronegative for HSV 2 antibodies before transplantation. Both developed the picture of primary infection after transplantation with disseminated vesicles throughout the body, which responded to antiviral treatment.

The isolates from the urine and vesicles of both patients were subjected to DNA "restriction endonuclease fragment length polymorphism analysis" (RFLP). Each strain of HSV 2, like each individual human being, has its unique RFLP pattern. The result of our test was that the RFLP of all the HSV isolates of the two individuals were identical. This suggests that the infection came from the same source, the common donor of the grafts in the two recipients.

We believe that before the donor's death, HSV 2 reactivated in his lumbar sacral ganglia. Active virus was reproduced, and virus traveled from the lumbar sacral ganglia via nerve fibers that innervate the bladder and ureter and contaminated the kidneys that were transplanted. This accounted for the HSV that was isolated from the recipients within days after transplantation. The virus then produced a primary infection in the recipients, which resulted in a disseminated picture ten and eighteen days after transplantation. This is not a common sequence of events involving HSV 2. It serves however as an example of how grafts may transmit viruses. Showing that the RFLP patterns of the viruses found in the two recipients were identical was another method of proving the origin of the virus in the donor[75].

Other Risk Factors of Herpesvirus Infection

From our discussion of herpesvirus infections after transplantation, it is apparent that the severity of herpesvirus infections varies a great deal with the circumstances and the type of transplantation. At one extreme, there may be only laboratory evidence of infection. At the other, there may be serious disease and even death.

One common denominator of all transplantation is the use of immuno-suppressive drugs. They act together with many other risk factors that could influence the severity of virus infections. To facilitate analysis each individual risk factor, we isolated the effects of immunosuppressive drugs, and two types of allograft reactions, that is, the host versus graft and the graft versus host reactions. To do this, we studied a group of patients out-side of transplantation and we embarked on experimental studies in mice.

Patients who are transplanted usually receive a combination of corticos-terioids, like prednisone, and one or more cytotoxic immunosuppressants such as azathioprine, cyclosporine or tacrolimus. We studied under the NIH program project funded in 1972, 131 patients who attended the rheumatol-ogy clinic of the University of Pittsbsurgh School of Medicine. They were on corticosteroids or on cytotoxic immunosuppressants, azathioprine or cyclophosphamide (similar to azathioprine in action), for connective tis-sue diseases such as rheumatoid arthritis, progressive systemic sclerosis or systemic lupus erythematosus. Urine and blood samples were collected for determining the presence of CMV and to measure antibody titers against CMV. Patients on corticosteroids alone had no evidence of CMV infection in the urine and blood. Those on cytotoxic immunosuppressants had signifi-cantly higher antibody titers against CMV compared to untreated controls although the proportion of seropositive patients was not changed. This was evidence that cytotoxic immunosuppressants activated latent CMV. Reactivation meant additional CMV was made, which stimulated more antibodies.

To nail down this hypothesis, John Dowling went to the rheumatology clinic of the University of Pittsburgh outpatient department, the Falk Clinic. With the help of the physicians in charge, he recruited fourteen patients who were on cytotoxic immunosuppressants for therapy. These drugs are at times beneficial in patients with severe rheumatological disorders such as systemic lupus erythematosus, rheumatoid arthritis or systemic sclerosis. They were followed prospectively for up to twenty months for development of CMV infection. Seven patients developed evidence of infection by a rise in antibody titer or by a positive urine culture for CMV. They were all seropositive before onset of therapy. There was only one seropositive patient who did not show signs of fresh infection. There were no primary infections. None of the infected patients had symptoms of illness.

This experiment shows that reactivation of latent infection of CMV occurs almost uniformly after administration of cytotoxic immunosuppressants. But these drugs alone were not enough to cause any CMV illness. Corticosteroids alone were inadequate to cause infection. Primary infection did not take place because there was no outside source for infection[76].

In order to test the reactions to a foreign graft on virus infection, we first established in mice a model of chronic mouse CMV infection.

Mice are the most popular animals for doing biological experiments. They are easy to breed and relatively cheap to maintain. An important reason is that so much is known about mice and so much has been done with mice. Work on mouse inbreeding and genetics, done primarily at the Bar Harbor Laboratories in Bar Harbor, Maine, has resulted in the production of many inbred strains, which are genetically identical. Inbred strains are essential for work on "allograft (or foreign graft) reactions", that is work on the reaction of the host against the graft (host versus graft reactions), and the reaction of the graft against the host.

Human CMV cannot infect animals, since it is restricted to humans. Mouse CMV is similar to, but not identical with, human CMV. Mouse CMV injected in the peritoneal cavity produces an asymptomatic chronic infection of indefinite duration in the salivary gland, with persistent viral replication. The virus can also be found in the spleen and kidney for six to nine weeks. The constancy of these manifestations allows us to measure the degree of chronic infection by measuring the amount of virus (its titer) in these organs.

To initiate a host versus graft reaction, a full thickness 15 mm skin graft was cut with a cork borer and sutured into a surgically prepared site on the mouse under anesthesia. An allograft or graft from a different breed of mouse showed signs of rejection, with signs of sloughing off and crusting after seven days. After ten days, the graft completely sloughed off. Control grafts from the same breed of mice showed no rejection and "took".

In the experiment, groups of mice were inoculated two to five weeks before skin transplantation to create a chronic CMV infection. They were divided into a group that received allografts and a group that received syngeneic (genetically identical) grafts. CMV titers, or the measurement of the amount of virus present, was uniformly higher in the tissues of kidney and spleen of mice with allografts from day three to day ten after transplantation. This experiment suggests that host versus graft reaction may be an additional risk factor for CMV disease[77]. It was carried out by Brian Wu, who was a research fellow of Chinese descent in our department. He was very capable in the laboratory.

A graft versus host reaction is when T lymphocytes from a graft recognize host cells as foreign and kill them. This reaction is a recognized serious complication of bone marrow transplantation. A graft versus host reaction can be created in a hybrid mouse, or a mouse whose two parents are of two different breeds (F1 hybrid), by injecting a graft of parental spleen cells, which consist largely of lymphocytes, into the hybrid mouse. The grafted cells survive in the hybrid mouse, but they recognize the host cells of the hybrid as foreign, and the grafted lymphocytes kill the host cells. Their being killed causes a graft versus host reaction.

To test the effect of graft versus host reaction on chronic CMV infection, we first created a chronic CMV infection in two groups of F1 hybrid mice. In one group, parental spleen cells were injected to create a host versus host reaction, and in the other, control spleen cells from syngeneic F1 strain were injected. Virus titers of kidney and spleen tissues were measured, as in the host versus graft experiment. CMV titers were uniformly elevated in mice that received the allografts, compared to control mice that received syngeneic grafts. This experiment explains why CMV disease is so much more severe in patients after bone marrow transplantation. They are subjected to an extra measure of immunosuppression arising from graft versus host reaction[78].

To Sum Up

By working out the mode of infection of a number of herpesviruses, we demonstrated the importance of latent viruses in transplant medicine.

Symptomatic disease is produced by these viruses in transplant recipients not only because of immunosuppressive drugs, especially the cytotoxic ones, but also because of allogeneic (genetically foreign) graft reactions, especially the graft versus host reactions in bone marrow transplantation. The type of transplantation may also contribute to susceptibility to infection in as yet unexplained ways.

We proved two maxims about three herpesviruses, CMV, herpes simplex and EBV. These latent viruses may be transmitted by the transplanted organ and produce significant disease during primary infection in recipients. Primary infections are more serious and present a greater risk than reactivation infections. These maxims have been extended in terms of hypotheses to other latent viruses that may complicate transplantation, such as HIV, hepatitis B, hepatitis C and herpesvirus types 6 and 8. They may also apply to as yet unknown viruses of human and animal organs and tissues that may be discovered to complicate transplantation in the future. Our studies have been a guide to the understanding of the pathogenesis and epidemiology of latent viruses after transplantation.

chapter

| 10 |

ACADEMIC MEDICINE

January 1, 1971 was the date of my inception as Chief of the Division of Infectious Diseases, announced by James J. Leonard, the new chair of the Department of Medicine after the retirement of Jack Myers. In the preparation for this new job, I was promoted to full Professor of Medicine in 1970.

I now had two full time jobs, the chair of the Department of Infectious Diseases and Microbiology in the Graduate School of Public Health, and this job in the School of Medicine. Part of the second job was being director of clinical microbiology in the Department of Pathology.

Technically, my primary appointment was in GSPH, but no one in the Department of Medicine, including myself, regarded my job there as secondary. In point of fact, the larger share of my salary was paid by the medical school.

With this appointment, the plan, "Infectious Disease Program at the University of Pittsburgh", which I submitted to the dean of the school of public health, Hershel Griffin, the dean of the medical school, Donald Medearis, and the vice chancellor for health affairs, Sargent Cheever, was being carried out.

The appointment represented the acme of my career in academic medicine.

I took over the suite of offices and laboratories at Scaife Hall vacated by my predecessor, Dr, A.I. Braude. Some of the offices and laboratories would be utilized by new faculty members I had to recruit. My other office and laboratories was at GSPH. The Division had one other faculty member, Chuck Craig, and a secretary, Betty Edwards. I started to move some of my books and paraphernalia to my new office. I used both offices, but I spent more time in my office at GSPH. This is not related to the amount of time spent on the affairs of the two schools. Rather it was because the office at GSPH was more pleasant. My office at Scaife was smaller, less well furnished, and there were wafts of odor from the animals in the animal room across the hall.

Academic medicine is a small part of the medical profession in the United States, but a highly respected part. It consists of the medically trained members of the faculties of American schools of medicine, schools of public health and medical research institutions. Advancement in medical schools is highly competitive. The requirement is excellence in clinical practice, research and teaching. One is judged by the medical school one attended, the hospital and service where one had clinical training, and one's research accomplishments. There are many medical schools in the United States, and their reputations vary. Big name schools help, but by and large, American medical schools are remarkably uniform in standards. Some schools produce more academics than others. This depends on the culture of the school, and their research record. The hospital in which one had his residency is important. After I graduated from Harvard Medical School there was a tremendous competition for residency slots. It was almost like applying to medical school again. The desired slots were in the internal medicine or surgical departments of the four Harvard teaching hospitals, Massachusetts General Hospital, Peter Bent Brigham Hospital (now the Brigham Women's Hospital) and the Harvard services at the Boston City Hospital. Slots in hospitals outside of Boston were generally less desirable. Only members of the class who were in the upper third of the class competed for these slots in Boston. I competed for a slot in internal medicine. I was accepted by the Harvard medical service at Boston City Hospital, and I considered myself very fortunate. Each service in each hospital had its own tradition, somewhat like a superior military outfit. At the Boston City Hospital, we prided ourselves in our ability to work independently and to be able to take care of emergencies and make do on a shoe string. I was well qualified for academic medicine in terms of the institutions where I received my medical education.

The usual medical graduate goes into practice after his residency training. But those who wished to pursue a career in a specialty, or those who wished to become a member of the academic medicine community must undergo a further phase of training. This phase was represented by what I did in the laboratories of Kass and Enders. This is called fellowship training.

Basically, fellowship training trains for a particular clinical specialty. My specialty was infectious diseases, and I was supposed to get clinical training in infectious diseases. However, during the three years I was an infectious disease fellow, about eighty percent of my time was spent in research in the laboratory and less time was spent in clinical medicine. As I was in a section at the Boston City Hospital that was active and well known in clinical medicine, I got most of my clinical training by daily contact with a staff that was constantly dealing with clinical matters. Since the fellowship is the last station of one's training before assuming a job, one's mentor as well as the reputation of one's fellowship is very important. Many fellowships are famous because of the institution they are in, or the mentor in charge. For those wishing to pursue a career in academic medicine, there are two possibilities after a fellowship. One can pursue a tenure track position in an academic medical institution, or one can pursue a non-tenure track research position in a university or in a commercial company. Graduates of Harvard Medical School usually applied for a tenure track position. Research positions are usually supported by grants or temporary funds, while tenure track positions are guaranteed by "solid" university funds. It is possible during good times and when research is flourishing for someone to shift from a non-tenured to a tenured position. But this is not usually possible.

One is considered for a tenure track position when there is a vacancy in the regular hierarchy involving teaching as well as research. One is evaluated for teaching and administrative ability besides research potential. The recommendation of one's mentor and the interview become important. It is important, for example, to be able to speak and write English fluently in order to be a good teacher or administrator. The process of medical education in the United States entails a long period of clinical training in medical schools and during residencies. This training requires presentation and discussion of cases day in and day out. So there is ample opportunity to be trained in public speaking during this time. This is the reason why American physicians usually have good communicative skills. This may also be the reason why in choosing potential candidates in academic medicine graduates of American medical schools may be inadvertently favored.

One's ability in teaching and administration does not change significantly with time but one's research accomplishments will. So a great deal of emphasis is placed on measuring scientific potential and accomplishments. There is an elaborate system to do this. One easy way is to enumerate one's publications in peer reviewed journals. Another method to measure excellence in research is election to elite research societies. Such societies consist of competitively elected members of the academic medical community. For example, when I chose Ed Kass as my mentor he was a young faculty member at Harvard who had just been elected to the American Society for Clinical Investigation, probably the most elite society of academic medicine. Members had to be elected before they were forty-three years old and the number elected each year was restricted to forty. By now the number has doubled. And the elitist aura around the society has dimmed a bit. Because of the high degree of competition for membership then, election to this society became an important criterion for measurement of quality. Unlike other criteria, this one is directly a reflection of the excellence of one's research. If one wants to become a professor of internal medicine or pediatrics, in the United States, it helps if one is a member of this society. This society is also called the "Young Turks". The Thorndike Memorial Laboratory at the Boston City Hospital, where I had my first fellowship, was at the time a hot bed for bright young investigators[79]. The best ones became members. They were usually elected five or six years into their investigative career, after they had made a significant impact in research, and after publication of a number of significant papers. Whenever one was elected, the entire staff of the laboratory joined in the jubilation.

I was elected in 1962 in Pittsburgh when I was thirty-five years old, and had already done research for five years. I was recommended by A.I. Braude and Maxwell Finland. Beyond the "Young Turks", there is another elite society called the Association of American Physicians, or the "Old Turks". Membership is also restricted by election. Most members are already members of the "Young Turks", and are already professors. Therefore it recognizes achievement rather than potential. I was elected in 1982. I was recommended by Maxwell Finland. Qualifications for this society are not to be scoffed at. A chair of medicine at Pitt was proposed three times before he was elected.

In 1959, I accepted the position of assistant professor at the University of Pittsburgh Graduate School of Public Health and School of Medicine after being recommended by Enders. I had joint appointments in the Department of Epidemiology and Microbiology, and the Department of Medicine. Three years later, I was promoted to associate professor. In 1965, at the age of thirty-eight, I was promoted to full professor at the Graduate School of Public Health. In 1970, I was promoted to full professor in the medical school. This relatively rapid course was made possible by my research accomplishments in interferon but it was facilitated by my academic background and training, my mentors and by my election to elite research societies.

Even though I intended to keep a strong foothold in clinical medicine, and insisted that I be given a joint appointment in the Department of Medicine when I first arrived in Pittsburgh in 1959, when I was appointed Chief of the Division of Infectious Diseases in 1971, I felt that my life had dramatically changed. I became administratively and academically responsible for a major medical subspecialty in the medical school, and in the teaching hospital. I was no longer a spectator and onlooker.

The biggest challenge facing me was that I did not feel confident that I could competently discharge my clinical responsibilities. I had, before me, the image of my predecessor, Abe Braude, a superb, world class clinician and bacteriologist, and an ideal chief of the division as well as chief of the microbiology laboratory. Every one in Pittsburgh who knew him was watching me carefully to see how well I could follow suit.

The fact is that I had not had any primary responsibility in clinical medicine since 1956, when I was a medical resident. I had an appointment in the departments of medicine at Harvard and Pitt all along; I attended the major clinical conferences, I saw patients in the outpatient department of Boston City Hospital, and in Pittsburgh I attended the weekly infectious disease rounds. At Pitt, the then chair of medicine, assigned me to be a "visit" (supervising, teaching physician) in medicine. This means going for three mornings a week, on "rounds" with the residents and medical students of a floor to take care of their patients one month at a time. I successfully passed the oral examinations of the Board of Internal Medicine on the first round and was certified. Despite all these activities in clinical medicine, I was essentially a full time researcher, and clinical medicine was a sideline. Things had now changed.

The main clinical responsibility of the Division was to respond to requests for consultations in infectious diseases in the two teaching hospitals, Presbyterian University and Veterans Administration (VA) hospitals. I had taken over the Division as almost the sole clinical faculty member. The sole survivor of the Braude team, Chuck Craig, had decided to transfer to the VA, so he took care of the primary responsibility for the consultations there. But I was left responsible for the main 600 bed hospital. We had no fellows in training, and assignments of medical residents to our Division were irregular.

For the first few months I had to answer the consults at "Presby" alone. This was a crushing load for me, as it is usually the full time job of one or two fellows or residents. But I did not complain, because I realized that the only way to regain my clinical competence was to assume actual clinical responsibility. To respond to the avalanche of consults was exactly the "crash" review course that I needed, even if there was no alternative.

A consultation was a request by the attending physician or resident for an expert opinion on the diagnosis or management of a patient with an infectious disease. Expertise requires that one must go beyond the obvious, which any physician can get in standard textbooks. Very often, I had to go to the library to look up the latest articles or reviews on the condition in question. Difficult cases might involve a more extensive literature search. At times, it was necessary to consult other experts in various parts of the country, but especially in Boston, by telephone. After expressing a written opinion in the chart, each case was followed visiting the patient, and writing follow up notes, until discharge. The following is from my diary.

> January 8, 1971. This is the first week I have assumed my new task at the medical school. It is a challenging task...I must master my day to day responsibilities. These are in 1), clinical infectious diseases, 2) clinical internal medicine and 3), laboratory diagnosis...This first week, I have already consulted on more than ten patients.

One vacuum left by the departure of Abe Braude was the supervision of the microbiological laboratory. This laboratory was administratively under the Department of Pathology. Initially, I asked Dr. Robert Yee, the bacteriologist in our department in the Graduate School of Public Health (GSPH), to help out. Within a few months I recruited Dr. Wie-shin Lee, a superb Ph.D. diagnostic bacteriologist, to supervise the laboratory. He stayed for about a year. To succeed him, I recruited Dr. William Pasculle, a doctoral graduate of Dr. Yee, who has remained supervisor until this day.

> September 3, 1971. I have been Chief of the Division of Infectious Diseases for eight months now. I believe I like this job... I have succeeded in the microbiology lab, where W.S. Lee is my supervisor, and where my authority is recognized by the pathologists.

To add to our faculty in the Division of Infectious Diseases, I recruited George J. Pazin, who was a University of Pittsburgh graduate, and a native of Johnstown, Pennsylvnia. He trained under Abe Braude, and was with him in La Jolla before he returned to Pittsburgh. At the same time, I recruited John N. Dowling for Presbyterian University Hospital. They occupied the rest of the offices and laboratory space we had at Scaife Hall. For the infectious disease section at the VA Hospital, which was also a part of the Division, I recruited Dr. Bosko Postic, who was already a faculty member in our department in the GSPH. He took the place of Chuck Craig, who left for the University of South Florida. Bosko was a close friend and a former student of mine. He was a native Serbian physician, who had trained in Boston under Max Finland and Ed Kass in Boston, where I first met him. He then came to our department in GSPH, where he obtained a doctoral degree, working on interferon and Sindbis virus under my supervision. Thus, he was broadly trained in medicine, epidemiology and virology. He finally found his métier in clinical medicine, leaving GSPH and being full time at the VA. Later at the VA, he was promoted to director of the medical service, and gave up his research interests in virology. He eventually transferred to the VA Hospital in Columbia, SC. In the eighties, he quit academic medicine altogether, and went into private practice in South Carolina, specializing in infectious diseases, with an emphasis on AIDS. In 1977, I recruited Victor L. Yu to head the VA infectious disease section, after Bosko became chief of medicine. Victor was a trainee of Tom Merigan's division of infectious diseases at Stanford. He is one of the most energetic and capable clinicians I know. He is also a prolific clinical researcher. Perhaps his best known accomplishment is his studies on legionellosis. He has remained in Pittsburgh, and developed a national reputation there. After arrival of new clinical faculty, consulting chores were equally divided among the professional staff throughout the year.

By 1974, we were able to attract infectious disease fellows for training. Within a year, we were to have one to three fellows all the time. They were internists who had completed their residencies in internal medicine, and would stay in training with us for two to three years. Their training consisted of clinical service on the consulting service, and research with a chosen member of the faculty. All the full time faculty had research projects. These were of course much strengthened by our association with the facilities at GSPH. Later, we were able to compete successfully for a traineeship program under the NIH, which required that we demonstrate excellence in clinical and research facilities for training of fellows.

Right from the beginning, my teaching activities in clinical medicine centered around the weekly infectious disease rounds each Wednesday afternoon. In the beginning, I organized these rounds personally. Later, the chore was shared by the rest of the faculty with the help of the fellows and residents on our service. This activity continued throughout the year. Interesting cases were presented from Presbyterian University and VA Hospitals by the responsible resident or fellow, and discussed by a faculty member or fellow. We invited the infectious disease groups at Montefiore (under Carl Norden, Fred Ruben and Ed Wing), and Children's Hospital (under Dick Michaels), to participate, and they would do so intermittently. Later we specified one Wednesday a month for the infectious disease specialists in private practice in Pittsburgh, and its vicinity to present cases. The specialty of infectious diseases in the United States developed *pari passu* with the advancement of my own medical career. It was not yet established when I was a fellow in the fifties. One might date its establishment with the founding of the Infectious Disease Society by Max Finland and his colleagues in 1962, and the subsequent establishment of the subspecialty Board of Infectious Diseases, as part of internal medicine. By the late 1970s, Pittsburgh and its environs had about ten practicing infectious disease specialists.

Following the tradition of Abe Braude, demonstrations from the microbiological laboratories were an integral part of these rounds. Therefore, our microbiology supervisors, first Dr. Lee and later Dr. Pasculle were also regular teachers in these rounds. They collected the laboratory materials for the demonstrations which followed the presentation of the cases, and they provided the microbiological information. When the cases required, there were also occasional demonstrations from the laboratories of the VA, Children's and Montefiore Hospitals. Rounds such as ours were offered by all the specialties in the hospital. Since they were an integral part of patient management, our infectious disease faculty, medical students (usually one to three third or fourth year students), residents and fellows in our specialty, and supervisors of the microbiological laboratories were expected to attend. I also offered these rounds as a course at GSPH, as many of our graduate students were interested in diagnostic microbiology, or infectious diseases. They came under the leadership of Dr. Yee, who was the director of our graduate program at GSPH, and who was a fervent supporter of our clinical program. He was always present during these rounds. He contributed to the discussions, especially concerning bacteriological points. Since these rounds were medically oriented, and full of jargon often incomprehensible to the layman, I would spend on the average about an hour each month to go over with participating graduate students, points that needed further explanation.

To provide the reader some idea of what was presented at these rounds, I will quote from my volume of records of the rounds from June 21, 1971 to June 20, 1973. This volume contains descriptions of all the cases presented, and the discussions of them by faculty members, residents, fellows or students. These discussions frequently included enlightening literature reviews. There were in the two years recorded 244 cases, or a mean of 2.35 cases per week. The volume consists of 162 single space typewritten pages. Examples given are cases presented in April, 1972.

April 5, 1972:   Salmonella Group D. A carrier in a 45-year-old female food handler.
Clostridium perfringes. Wound infection in a 88-year-old woman after pinning of intertrochanteric fracture.

April 12, 1972:   Taenia saginata (beef tapeworm). A 43-year-old woman who has traveled to the Far East and who saw "tissue" in bowel movement.
Visceral larval migrans. 21-month-old boy from Children's Hospital with convulsions due to systemic dissemination of larval forms of Toxicara.
Miliary tuberculosis. Male patient with signs of "flu" and t.b. meningitis.

April 19, 1972:   Genital herpes. A 20-year-old black married male.
Fever and Herpes simplex infection. 13-year-old renal transplant recipient.
Staphylococcal enteritis. 23-year-old male with ulcerative colitis, developed "flu" like syndrome. After treatment with penicillin, developed enteritis. Staphylococcus aureus was isolated from stool.

April 26, 1972:   Tuberculosis of the kidney. 65-year-old black lady with "fever of unknown origin".
Pneumococcal meningitis (2 cases). 35-year-old woman with history of alcoholism and a 25-year-old woman with chronic bronchitis. Both patients died.

Our contribution to the understanding and management of viral infections after transplantation was an important chapter in my career. Under my tutelage, Pitt produced a number of clinical leaders in transplant infectious diseases, such as Steve Dummer, Shimon Kusne, Penney Williams and Nina Singh. They are a new breed of specialists whose specialty is transplantation infectious disease. Because of my clinical expertise in this area, I was voted one of the "Best Doctors of America" in 1992 and 1993. To me the exciting part was exploring the new frontiers.

Teaching

It might be apparent to the reader that the common denominator of my research and clinical activities was my teaching. In retrospect, I rejected several opportunities that were presented during my lifetime for a career in a research institute separate from a university because there was no opportunity for teaching. As can be seen from the account of my clinical activities above, they were intimately related to teaching. I considered once in my career at the University of Pittsburgh the possibility of leaving the university for private practice. This was at a time of my discouragement around 1970, when I failed to get a renewal of my NIH research grant in interferon. I did not take long to decide that I would stick it out in academic medicine and remain in teaching and research.

Teaching can take different forms. To me the most enjoyable form was one to one teaching, as in the guidance of a doctoral or fellow's dissertation. An equally enjoyable form was small group teaching, as during clinical rounds in a hospital, or in small group seminars in a lecture room. This type of teaching is conducive to presenting a particular point of view, rather than simple recitation of facts, or summary of an accepted principle.

My experience in teaching cannot be discussed separately from my ability to speak, write, and do research. Ability to do these three things well makes for good teaching. As far as lecturing is concerned, I already pointed out that clinical training and experience as a researcher trained me in public speaking, as well as in writing. Presenting or writing a case report, or a piece of research requires the ability to organize factual material, and to make sense of the subject matter. My first attempt at serious writing was my senior thesis at college. At that point, I did not let my colleagues see my work. This was a reluctance that must be overcome, because critiques of outsiders are an essential requirement for improvement in writing. At Stanford University, I took an introductory graduate course in political science called "Political Science 400", given by Professor Charles Fairman, who was a famous constitutional lawyer. He stressed the essentials of good writing, emphasizing the need to think clearly and the importance of precision and conciseness. There is a saying, "Easy reading is curs't hard writing, easy writing is curs't hard reading." These ideas may be simple, but a lifetime of effort is required to maintain the quality of writing.

Effective lecturing has the same requirements. Actually, the ability to teach may be acquired from any serious productive discussion, because it often requires precise, clear thinking and lucid exposition. I learned how to present a complex problem by reporting on difficult cases in clinical medicine. By discussing with students how to approach and solve a research problem also teaches one how to teach. At the University of Pittsburgh, I acquired the ability to explain a complex problem in simple laymen's language. I had to do this when I was responsible for an introductory course in medical biology for entering graduate students at GSPH, who never had a course in biology. After learning how to do this, I was able later to write articles for laymen in newspapers and magazines in Taiwan, even though it was in Chinese.

What does research have to do with good teaching? This is a point frequently debated, particularly by teachers in higher education. Surely, there are good teachers who do no research. In fact I know one teacher in the Graduate School of Public Health, a few years younger than me, with almost the same type of educational background as mine, who upon being appointed assistant professor, and beginning his career as a teacher of epidemiology, stated flatly that all he wanted to do was to teach. He let his superiors know that he had no interest in research. Indeed, he was a fine teacher, and he later received the "Golden Apple", awarded by students for excellence in teaching. Parenthetically, this is an award I never got.

When I gave the introductory course in human biology at the Graduate School of Public Health for non-medically trained students, who had no background in biology, in the beginning I was challenged to explain complex matters simply and lucidly, and I enjoyed it. But that lasted only for a few years. I then was "burned out". I was happy to turn it to some one else. I could not have followed my epidemiologist friend, who to this day, forty years later, is still lecturing and teaching his basic course.

One of the excitements for me about teaching is presenting new material. This is not possible in rote lecturing. It is best illustrated by the inexhaustible source of inspiration one gets in clinical case teaching. Bedside teaching is teaching by going over a new case every time. Whenever one discusses a new case in medicine, there is always something new, something that is not exactly in the text book. The challenge of a new case is making a diagnosis, prescribing a made to order program of treatment, or the challenge of disposition. In short, it is teaching by solving a problem.

When I was a student at Harvard Medical School, and for the first years after I came to Pittsburgh, I participated in a clinical teaching-learning exercise, called the "CPC" (clinical pathological conference). The discussant has to discuss an unknown case, the diagnosis and answer of which is known only to the pathologist, who reveals it at the end of the discussion.

These conferences are held in a large auditorium, and the participants are members of all the clinical departments. That by the way is the reason why this very useful exercise is no longer in vogue in medical schools. Conferences attended by all clinicians have now become almost extinct. They have been supplanted by conferences of narrow specialists. But in the early days, there was still interest in general medicine. In Pittsburgh, a frequent discusser was Jack Myers, the founding chair of the Department of Medicine. Jack was the last "Renaissance" man among the chairs of medicine that I knew at Pitt. He aspired to know, professed to know and actually did know most everything about general medicine. He retired from the chair to devote himself full time to the task of incorporating his personal base of knowledge in a soft ware he worked out called "The Internist". In this age of specialization, no one aspires to know everything any more, and no single person would dare to produce such a program. In any case, Jack was often asked to discuss cases, and he never refused. Nor was he embarrassed by missing a diagnosis, even if there was a certain amount of "Schadenfreude" among the audience when he failed, because he was an almost arrogantly self confident man. But Jack always gave a brilliant discussion, which bristled with "pearls" and elegant logic.

I also gave a few CPC's. The amount of research I did in working up the cases given to me was tantamount to small research projects. One time I was given an obscure case of encephalitis with no microbial diagnosis. From my search of the literature, I came up with the diagnosis of "inclusion body encephalitis", a diagnosis that is no longer made. But I knew from my virology that there are only a few viruses that produce inclusion bodies in the cells of the brain, and one prominent one is herpes simplex. Indeed herpes simplex was isolated at autopsy. I became famous in our small medical community because of this feat. This is the type of teaching-learning that I liked.

So the relevance of research to teaching is basic to me. Which is the same thing; it is the task of solving a problem. Each research problem is different. Each solution is unique. This is why I like teaching individuals, because they either have a unique research problem, or they have a learning problem that I have to address. I am inspired by new solutions, new insights, new approaches, and I like to transmit these in my teaching rather than rote knowledge. I am inspired by the search of discovery.

Perhaps there is a place for both types of teachers, teachers like my epidemiologist-friend, who are more like high school teachers, and the type inspired by research.

Students and fellows with whom I did research work were closest to me. In more than fifty years of academic life involving the writing of more than 280 scientific papers, graduate students, research fellows and colleagues were frequently co-authors. All graduate students, whether working for a masters or a doctoral degree, had to write a thesis. Doctoral theses frequently ended up as publications. The same was true of research done with our infectious disease fellows. My habit over the years was to spend every Wednesday morning discussing research with students and fellows. We covered the problems of all who were involved with a research project. We reviewed the work done and we planned for new work to be done. The method of approach as well as the organization of scientific papers were also discussed at these meetings.

Sabbaticals

There is a system of granting sabbatical leaves to tenured faculty members in many American universities. Theoretically, a sabbatical leave may be applied for every seven years. It can be of one year's duration, but one is only given a salary for six months. One may apply to outside foundations for additional support or for travel.

During my forty years as a professor, my two sabbaticals were not only unique cultural and academic experiences for me, but they were also unique for the life of my family. My first sabbatical was in Germany. Ever since the end of World War II, I had looked forward to a revisit of Germany and Austria, countries where I had spent some years when I was a boy. The opportunity came in 1965, when I was awarded a senior Fulbright fellowship for a 10-month sabbatical at the Max Planck Institute of Virology in Tübingen in southwest Germany. At that time, Germany and Austria had recovered completely from the ravages of World War II, and were in many ways similar to the countries I knew when I was a boy. I had the opportunity to improve my German, a language that I had struggled to keep up all these years. My spoken and written German reached the point where I was giving scientific talks in German, and before I left Germany, I was even considered for a professorship in a German university.

While in Europe, we also took the opportunity to visit old friends in Vienna. One of them was Mrs. Elizabeth Murray, a friend of my father's, whose address in the 18th District I had remembered all these years. Her son, Kurt, had since immigrated to the United States, and we had resumed

Carol, Bettie (age 9) and John (age 6), photo in their joint passport, on my sabbatical in Gemany (1965–1966).

contact there. Mrs. Murray and her daughter, Lilly, arranged for a reunion with us at the South Railway Hotel in Semmering, near Vienna, one of our old haunts. We also saw Otto Pelz, whose family owned a shoe store. One day, I thought I would look up a classmate, Karl Hauswirth. Since I did not have his address, I just looked up his name in the phonebook, and I found it there. He was a practicing psychologist, and I visited with him and his young wife where we had a delightful hour at his office — reminiscing about our primary school days at the Rudolf Steiner School. Not only I but the rest of the family was also thoroughly inundated in German culture. Our 9-year-old daughter, Bettie, entered the 3rd grade of the local Waldorf School in Tübingen. This was quite similar to the Rudolf Steiner School that I had entered as a child in Vienna, so I was quite familiar with its curriculum. Bettie is quite adept at learning foreign languages, and was speaking German fluently within months. We enrolled our 6-year-old son, John, in kindergarten. He did not speak a word of German for three whole months. Then one day, he broke forth in fluent German, not high German, but the local Swabian dialect! Like me in Turkey, he picked it up playing with the kids. This is further evidence of the ease with which children pick up foreign languages. My wife, Carol, became active in the local American club, and made many American and German friends. She was able to get along in broken German, which was sufficient for shopping and among friends. Years later, both children would frequently cite this experience abroad as one of their most enjoyable and profitable times in their childhood.

The Fulbright Commission was very active in arranging my extracurricular activities. They distributed my name among many European universities such that in the course of a year, besides universities in Germany, I was invited to give lectures at many other European institutions. For some reason, I was particularly popular in Italy, where we visited the universities in Siena, Perugia, Naples, and Polermo in Sicily. I was also invited to speak at universities in Madrid, Spain, in Bratislava, Czechoslovakia, and in Sheffield and Aberdeen, in the United Kingdom.

For my second sabbatical (1978–1979), I applied to the Josiah Macy Foundation for support, and the John Curtin School of Medical Research, Australian National University in Canberra for research. My wife, Carol, and our 19-year-old son, John, and I lived in a spacious house on the university campus reserved for professors. Australia, like the United States, belongs to the "new world". It is a continent by itself, gigantic in length and breadth, but with a population only the size of Taiwan. My first impression of Australia was that, even in the heavily populated southeastern portion, Australia was a lonely place. On the other hand, since Australia was settled in the early part of the 19th century, it has in 200 years developed into an advanced, technologically sophisticated country worthy of our admiration. The accomplishments of Australia are undoubtedly related to her being an English speaking country, close to the United Kingdom, and more recently to the United States. Thus, similarity of language and culture is much more important than geography in the development of countries.

My research topic was CMV cellular immunity, about which there was a lot presumed, but relatively little known. I started to work in Australia like a research fellow when I was in Boston. I did everything in the laboratory by myself, with help of half of one technician. Before I could start to do any meaningful experiments in cytotoxic immunity against murine CMV, I had to establish the components of my experimental system from scratch. First, I had to establish a mouse cell culture system, so that I could grow murine CMV. I had to establish a CMV titration system by plaque titration. The next job was to establish the cytotoxic system. We had to generate cytotoxic T lymphocytes, which are responsible for killing CMV infected cells in cytotoxic cellular immunity. Such cells could be obtained by isolating lymphocytes from mouse spleens ten days after infection with CMV. The cell number could be expanded *in vitro* by incubation with infected cells. The next requirement of the cytotoxic system was to measure the cytotoxic or cell killing effect. Cytotoxic T-cells are incubated with target cells infected by CMV, and labeled with chromium 51. Killing by specific cytotoxic T-cells is measured by release of the label from target cells into medium using a radioactivity counting machine. The amount of chromium 51 released is the measure of cell death or cytotoxicity. After establishment of these complex systems, I was finally able to undertake the definitive experiments. The idea was simple. We compared CMV virus replication in mice which were treated with CMV specific cytotoxic cells, with replication in mice that were not treated. A few days after infection with CMV, the spleens from both treated and untreated mice were isolated and the titers of CMV were compared. We found that the titer of CMV decreased by treatment, or the number of specific cytotoxic cells inoculated. Therefore cytotoxic immunity had a definite ameliorating role in cytomegalovirus infection. In a relatively short period of ten months of work, I was able to complete the entire set of complicated research tasks and finish two complete papers[80,81]. When I reported on my research at a departmental meeting at the end of my sabbatical, the head of the department, Dr. Gorden Ada, was deeply impressed. For me, this sabbatical leave, although extremely busy, was a welcome change from my usual administrative routines.

Thus, the two sabbaticals not only advanced my scientific education and accomplishments, but they introduced to my family the diversity of Western culture and history.

chapter

| 11 |

THE UPS AND DOWNS OF A DEPARTMENT

On March 25, 1996, Dean Donald Mattison of the Graduate School of Public Health (GSPH) announced at a meeting he asked me, as department chair to call, that the Department of Infectious Diseases and Microbiology (IDM) would be dissolved. It was to be incorporated into the Department of Environmental and Occupational Health (EOH) chaired by Dr. Herbert Rosenkranz.

This was the apparent solution of a long standing problem: the problem of my successor as chair of the Department of IDM after my retirement.

In order to understand the background of this difficulty, one should first consider the nature of the teaching programs in American schools of public health. What is the role of basic disciplinary sciences, such as microbiology, in schools of public health?

What and how to teach is a recurring problem of the faculty in schools of public health. It has been constantly discussed during my more than forty years as a faculty member at GSPH. The answer would seem to be inherent in the mission and objectives of the school. That should determine the types of specialists which it wishes to train and the academic disciplines required to do so. But it is not clear cut. When I first came to the school, perhaps being influenced by my own interest in basic research developed during my traineeship in the laboratories of John Enders, I fought hard to establish a program of microbiology with strict attention to disciplinary excellence. Not that I was not interested in problems of health, for I was a dyed-in-the-wool medical academician. But I felt that basic science was just as important in a school of public health as in a medical school. I felt that excellence in microbiology was essential for research in infectious diseases of individuals, as well of the public. This view of basic science was not the consensus in schools of public health. Some felt that there must be a clear distinction between schools of medicine and schools of public health. Basic science was pursued in medical schools, and schools of public health should pursue something more practical, more immediately applicable to public health. There were a few at GSPH who shared my point of view. One was C.C. Li, the world class geneticist. His concept of basic science, specifically related to human genetics and biostatistics, was as rigorous as mine regarding microbiology and infectious diseases. But C.C. remained a pure scholar throughout his life, and eschewed public debates and advocacy. Our views notwithstanding, what was and is the accepted view of academic disciplines in schools of public health? I feel the views have changed over the years. In order to understand the situation, it is necessary to review some of my history and GSPH's history.

My precise career objectives were still unclear when I accepted the offer of the University of Pittsburgh to be an assistant professor in its Graduate School of Public Health and School of Medicine in 1959, at the age of 32. I wanted to be in academic medicine, which traditionally involves teaching, research and clinical practice. With these priorities, it may seem strange that I accepted a primary appointment in the Department of Epidemiology and Microbiology of a school of public health. In retrospect, I was influenced by the exceptionally warm welcome accorded to me during my first visit by Dr. William McD. Hammon, the department chair, and his staff, and the other full professor in the department, Horace M. Gezon. The department also had other connections with Boston and Harvard, where I had studied for eight years. I graduated from Harvard Medical School, had an internship and residency at the Harvard medical service at the Boston City Hospital, and completed a three year fellowship with Ed Kass and John Enders at Harvard Medical School. Hammon was also a previous student of John F. Enders, the Nobelist and virologist who was recommending me and who exerted the greatest influence on me in my scientific development. The person on the faculty whom I was replacing, Francis S. Cheever, was a well-known Bostonian and Harvard physician-virologist, also well known to Enders. Cheever was being promoted to become the dean of the medical school at Pitt. In addition, my priority at the time seemed to be dominated by research. I chose the school of public health at Pitt because there I had available adequate research space and time, more than I could be offered

Monto Ho succceeds Willliam McD. Hammon as Chair, Department of Epidemiology and Microbiology, Graduate School of Public Health, Unversity of Pittsburgh (1969–1972).

in a school of medicine. I did insist on having a joint appointment in the Department of Medicine because I wanted to be a *bona fide* member of the internal medicine faculty. When I arrived, I had already acquired a research grant from the NIH. Support by the National Institutes of Health gives the principal investigator a great deal of independence. This released me from asking for research resources from the University. In this respect I was very fortunate, because from 1959 until 1997 when I retired, I had been almost continuously supported in research by the NIH. Support from this source is not only important financially, but in the United States, it is also a sign of research distinction.

My insistence on a joint appointment in the Department of Medicine had profound effects. Their head of infectious diseases was the well-known A. I. Braude, whose insistence on quality was well known. I was very fortunate in having his support for my activities and development. In 1962, it was he who recommended me for membership to the competitive elite Young Turks (American Society for Clinical Investigation), with an important endorsement from my previous chief in Boston, Max Finland. Because of Braude, the whole Department of Medicine considered me one of their own. Therefore in 1968, when Braude took an equivalent job at the University of California in San Diego and moved away from Pittsburgh, the chair of medicine at Pitt, Jack Myers, offered me Braude's position. But he insisted that I become full time in medicine and not in the school of public health. It turned out that just at that time my chief in the school of public health, Bill Hammon, announced his retirement. The school promptly offered me the position of the chairmanship of the Department of Epidemiology and Microbiology at the school of public health without any outside search. I accepted their offer but did not formally reject Myers' offer.

After being in my new position in the school of public health for about half a year, I felt an unforeseen pressure. In the beginning, even though I had no background or expertise in epidemiology, I thought that as a chairman of the department covering both epidemiology and microbiology, but being primarily interested in microbiology and infectious diseases, I could still take care of two areas by recruiting a strong faculty member in epidemiology, especially in chronic diseases. This assumption turned out to be incorrect, because it did not take into account the fact that epidemiology is a pivotal discipline in a school of public health, and cannot be led by a part time person. In 1969, I participated in the famous summer program of epidemiology courses at the University of Minnesota. This was intended to give me a review of epidemiology as a discipline. But more importantly I had the opportunity to get to know prominent members of the epidemiology community. I met a young man by the name of Lewis H. Kuller, an upcoming chronic disease epidemiologist, who was giving the course on chronic disease epidemiology. After listening to his lectures, I discussed with him the possibility of his joining the faculty at Pitt. He seemed to be interested and I invited him to visit Pitt later that year. He came and we offered him a professorship in epidemiology. His response seemed to be positive but he remained indefinite. At that time, Kuller was a young associate professor at the Johns Hopkins School of Public Health. One day, I was a guest lecturer at his department after having been invited by Abraham M. Lilienfeld, a well known chronic disease epidemiologist who was both the mentor and superior of Kuller. I asked Dr. Lilienfeld what I had to do to get Kuller to come to Pittsburgh. He replied frankly that what Kuller wanted was not just a professorship of epidemiology but the chairmanship of the department. I answered that if that was all there was to it, there would be no problem. On returning to the University of Pittsburgh, and after consultation with our then dean, Herschel E. Griffin, we decided to offer Kuller the chair of our department. I relinquished the chairmanship but remained professor of microbiology in the department. This offer and arrangement was accepted by Kuller. After thirty years as chair, Dr. Kuller retired in 2002. He established one of the finest departments of epidemiology in American schools of public health. This event describes something I did for Pitt at the cost of my own position.

The problem of epidemiology was satisfactorily solved after the arrival of Kuller, only to be replaced by the problem of microbiology. To Kuller, microbiology was but a method for the study of epidemiology, and he did not believe it was necessary to have an independent program for it. My views of microbiology were different. Firstly, my starting point in microbiology is an interest in infectious diseases, and in this respect Kuller's and my views were not too different. But secondly, my interests in microbiology were developed when I was a fellow in Kass' and Enders' laboratories. They developed my interest in basic mechanisms. That is why I delved deeply in interferon research. In order to cultivate such an interest, I needed a basic science microbiology program. Even though I was in a school of public health, all my masters and doctoral students developed a sound foundation in basic microbiology. My way of running microbiology was thus contrary to Kuller's. It was not possible for both of us to have our ways in one department. We presented our different views to Dean Griffin. My views were supported by Dean Griffin. He decided in my favor by proposing to the school and the university in 1974 that a new and separate Department of Infectious Diseases and Microbiology separate from Kuller's department be founded. I was appointed chair.

Infectious diseases and microbiology were not primary or major disciplines in schools of public health in the United States during the seventies. At that time, contrary to perceptions nowadays, infectious diseases were thought to be problems of the past. They had already been solved or methods to solve them were apparent to all. It was no longer necessary to study them academically in depth. There were relatively few substantial programs in microbiology in the twenty or so schools of public health. Even so, I felt at the time that I had an unusual opportunity to develop a major comprehensive program at the University of Pittsburgh. I had a foot hold in both the school of public health and in the school of medicine. I could combine the resources of two different schools in order to achieve a unique program.

I took advantage of existing arrangements and suggested that they be combined under one leadership. I authored a proposal called "Infectious Disease Program at the University of Pittsburgh". This proposal was given to the dean of the medical school, Donald Medearis, the dean of the school of public health, Hershel Griffin, and the vice chancellor for health affairs, Sargent Cheever, and they accepted it. The plan called for my being the chief of the Division of Infectious Diseases in the school of medicine as well as the Director of the Microbiology Laboratory at the main teaching hospital, the Presbyterian University Hospital. Both of these appointments were in the school of medicine, one was in the Department of Medicine and the other in the Department of Pathology. My third appointment was to be the chairman of the Department of Infectious Diseases and Microbiology in the school of public health. The teaching and clinical activities of the three units continued as they were, but the research activities of the three units would be pursued under one leadership.

This new arrangement was novel for the three units that I now headed. It turned out that with the departure of Abe Braude, there was only one remaining faculty member left in infectious diseases and diagnostic microbiology in the medical school, and he soon left. So I had a recruiting job ahead of me. But for the faculty of the Department of Infectious Diseases and Microbiology (GSPH) this change marked not only in the nature of the leadership but also in departmental direction. The primary interest of Dr. Hammon, the previous chair, had been in arbovirus epidemiology, and he had essentially no relationship with the school of medicine. I hoped to develop the graduate program at GSPH by utilizing the new clinical connections in infectious diseases and diagnostic microbiology. One could envision many new opportunities for research by master's and doctoral candidates. Faculty members who were within my own research areas, such John Armstrong, Bosko Postic, and Mary Kay Breinig were naturally agreeable to this change. But I probably would not have been so bold as to take on these three jobs were it not for the encouragement and assistance of one senior faculty member, Robert B. Yee.

Bob is a native Pittsburgher and a graduate of the University of Pittsburgh. He obtained his doctoral degree working with Shigella with Horace Gezon, a previous faculty member. First Bob assured me that he would take care of most of my administrative work at GSPH. He was appointed associate chair of the department to facilitate his doing that. Second, he took on the job of running the graduate training program. For thirty years, he was "godfather" not only of his own students, but all our graduate students, including mine. Bob was able to do such a superb job because he was so devoted and effective. Third, Bob is an accomplished bacteriologist, an area outside of my main interest. He filled an important need in the department since most of the rest of us were interested primarily in viruses. He also helped me supervise the diagnostic laboratory in the hospital in the beginning. In everything he did, Bob exhibited a devotion and selflessness which is unique. He was in the department when I came to Pittsburgh, and he was still in the department, helping out, even after his and my retirement. His primary concern was always the department. Without him, I could not have embarked on this new adventure.

This administrative arrangement continued for twenty-three years. It might be considered as a personal achievement at least in the administrative and research sense. In looking back, I see that this plan became the backdrop of almost my entire academic career. It represented the broad scope of my interests in infectious diseases, which covered its clinical activities, diagnostic microbiology as well as its epidemiological and public health aspects. Only under this unusual arrangement could all these interests be developed by one person.

Lest my personal experience appear prejudicial, I might cite the prevailing view on the role of basic sciences in a school of public health as illustrated by the history of the Department of Biochemistry and Nutrition at GSPH. When Thomas Parran founded the Graduate School of Public Health in 1948, one of the five original departments he created was the Department of Biochemistry and Nutrition. He recruited as chair a young and brilliant physician and basic biochemist, Robert E. Olson. Dr. Olson successfully developed a public health nutrition program as well as a basic biochemistry program which trained masters and Ph.D. students. In 1965, he resigned and left the university. Under his successor, Richard Abrams, a non-physician basic scientist, the nutritional component of its programs was phased out and basic biochemistry became dominant. In 1966, the department was rated by an external committee of GSPH as "unrelated to the mainstream of public health" (p. 156)[82]. In 1969, the department separated from the school of public health and it became the Department of Biochemistry in the School of Arts and Sciences. The entire department was eliminated without a replacement. The lack of even a champion for nutrition, the public health aspect of biochemistry, is ascribable to the lack of a strong proponent of chronic disease epidemiology, who would have been interested in nutrition. The chair of the epidemiology at the time, my predecessor, William McD. Hammon, was a pure infectious disease epidemiologist of the classical type, without a whit of interest or concern about non-infectious diseases. This experience tells us that basic disciplinary studies are tolerated in schools of public health as long as they are led by strong leaders and if they demonstrate relevance to public health.

Whether a department or program exists in a school of public health is also determined by what is considered important in public health. The importance attached to infectious diseases and microbiology in schools of public health has undergone vast swings. It was the preoccupation of American schools of public health around the time the School of Hygiene and Public Health at Johns Hopkins University was founded in 1918. By the 1970s, more than half of twenty or so schools had no department or program in infectious diseases. Only the larger and more prestigious schools, such as Harvard and Johns Hopkins, had academically strong ones. The major problems of infectious diseases were thought to be resolved, at least theoretically if not practically in the real world, by achievements in environmental sanitation, a working public health infrastructure, potent vaccines and antibiotics. The eradication of small pox in 1979[83] and the availability of a giant armamentarium of antibiotics against practically all bacterial infections contributed to the general feeling of security against infectious diseases.

That this was a false perception was clearly stated by the awakening call about the "emerging" infectious diseases in an Institute of Medicine paper led by Joshua Lederberg in 1992[84]. Public realization that the threat of infectious diseases was greater than ever also came with the discovery of AIDS in 1981, a heretofore unknown disease of man. The subsequent world pandemic of this viral disease is rampant and raging throughout the world even as I write more than twenty years later. The disease has not yet been prevented by an effective vaccine or cured by potent antivirals, despite a prolonged worldwide scientific effort to achieve these goals. It seems to me that if the existence of a separate Department of Infectious Diseases and Microbiology in a prestigious school of public health was already justified in 1970, it was even more so in the eighties.

The First Threat to the Department

In the summer of 1984, when I was overseeing a clinical trial of recombinant interferon on hemorrhagic fever with renal syndrome in Wuhan, China with my previous fellow, Dr. Xien Gui[59], I received a telegram from Bob Yee, who was acting chair of the department of IDM in my absence. The new vice chancellor for health affairs, Thomas Detre, was considering the dissolution of the Department and incorporating it with the Department of Microbiology in the medical school.

I was alarmed. First of all, it is unusual these days to receive any telegram. The telephone and fax machines have taken its place. Secondly, I was afraid that in my absence the intended action would become a *fait accompli* before my return. I decided to terminate my trip immediately and return to Pittsburgh to talk to Tom.

Tom Detre and his wife Catharine are both eminent medical investigators who emigrated from Hungary. Tom was recruited from the Department of Psychiatry at Yale University to become professor and chair of Pitt's Department of Psychiatry and superintendent of its internationally known state supported teaching hospital, WPIC (Western Psychiatric Institute and Clinic). When I first arrived in Pittsburgh, WPIC was one of the strongholds of psychoanalysis in the United States. Within a few years after Tom's arrival, psychoanalysis was effaced from the scene and replaced by a strong faculty of non-analytic psychiatrists who stressed drug therapy, and were devoted to organic and epidemiological research. His success in creating a new department was such that he was appointed vice chancellor of health affairs. His jurisdiction was the five health related professional schools; medicine, public health, dentistry, nursing and health related professions. During his tenure of almost twenty years as one of the most successful vice chancellors of health in Pitt's history, he transformed the University of Pittsburgh Medical Center (UPMC) to a nationally recognized quality health center, and Pitt's medical school to one among a dozen or so top medical schools in the United States.

As soon as I returned to Pittsburgh, I made an appointment to see Tom. He did not move to the medical school at Scaife Hall after his promotion but remained at his warm and spacious office at WPIC, lined with book cases and rugs and comfortable chairs. I had met Tom previously when I attended the monthly executive committee meetings of the medical school as an elected member of the Department of Medicine. Tom is a relaxed, confidant person of medium stature. He stood up to greet me with old world courtesy. He speaks slowly, deliberately and precisely with a characteristically melodious Hungarian accent. I was meeting him for the first time as chair of the Department of Infectious Diseases and Microbiology in the Graduate School of Public Health.

Our conversation began in a light vein. I had just returned from China, which had become a popular and favorite place for American academics to visit since China was "opened" to the outside world by President Nixon and Henry Kissinger.

I asked about his plans for the Department of IDM. He said, "Your department can join a larger group in the medical school and increase its critical mass. Merger is a possibility, but a final decision has not been made."

I took the opportunity to present my view. I said, "There are four departments at the University of Pittsburgh with the word 'microbiology' in its name; the Department of Microbiology in the medical school, the Department of Microbiology in the dental school, the Department of Microbiology and Biophysics in the Faculty of Arts and Sciences, and the Department of Infectious Diseases and Microbiology in the Graduate School of Public Health. The first three departments are all staffed by Ph.D microbiologists even though they are in different schools. Their research is in basic microbiology, often with a molecular twist. The Department of IDM is different. It alone deals with infectious diseases in three unique ways; in its clinical aspects, its clinical microbiological aspects and in its epidemiological aspects. We are able to do this because I am also chief of the Division of Infectious Diseases and Director of the Clinical Microbiology Laboratory in the medical school, and we are situated in a school of public health with strength in epidemiology. Our teaching and research activities are distinct from the other three departments."

He accepted my argument, and we survived for the next 15 years until my retirement!

In the late 1980s, I presented to Tom a plan to enlarge the virology program in the Medical Center by adding one additional faculty member strong in molecular virology in each of the three units under my direction, that is the clinical microbiology program in the Department of Pathology, the Division of Infectious Diseases of Department of Medicine and Department of IDM at GSPH. Tom was enthusiastic. He called together my immediate superiors in the three units under my direction, two department heads and one dean. They listened to my proposal and agreed to it within fifteen minutes. It added significantly to our strength in molecular virology. Thus Tom and I had an excellent history of collaboration.

Different Fates of My Three Jobs

My professional career at the University of Pittsburgh encompassed three jobs. They combined different aspects of my interests in infectious diseases and microbiology. The geography of the three different departments to which my jobs belonged, and the leadership of the University Health Center make such amalgamation possible. With my imminent retirement, the fate of each of these three jobs underwent different devolutions. No thought was given to a continued merger, for each of the three jobs had gotten too big and complex.

In January, 1971, I took over the Division of Infectious Diseases in the Department of Medicine as chief, and I retired twenty-one years and six months later in July 1992 at the age of 65. George Bernier, Dean of Medicine, appointed Ed Wing, a professor of medicine in our Division of Infectious Diseases as my successor. When I succeeded Abe Braude as chief of the Division in 1971, I was its sole faculty member, and we had no fellows. When I retired, we had a faculty of thirteen, and six fellows. Our clinical responsibilities were distributed in the three teaching hospitals of University of Pittsburgh Medical Center (UPMC), the Presbyterian, Montefiore and Veterans Administration Hospitals. The Division has continued to expand and thrive after John Mellors was appointed in 1999 as chief. John also rose from our ranks.

The Clinical Microbiology Laboratories, located primarily at Presbyterian University Hospital, has had the smoothest transition after my retirement from being Director of Microbiology in 1994. This is because the two individuals I recruited in 1976 and 1978, Bill Pasculle and Chuck Rinaldo, to supervise the bacteriology and virology aspects of the laboratory, were and still are in place.

In contrast to the smooth transition following my retirement from my two jobs in the School of Medicine, the search for my successor and chair of the Department of Infectious Diseases and Microbiology (IDM) at the Graduate School of Public Health (GSPH) met with some controversy. Unlike my two other jobs in the School of Medicine, which were well established and had their own traditions, the Department of IDM at GSPH was essentially my creation. The department was sequestered from the Department of Epidemiology and Microbiology with my guidance in 1970. It had seven faculty members at my retirement. With my impending retirement, its future prospects became a debatable issue.

In December, 1992, with my impending retirement, Tom let it be known through our Dean, Dr. Don Mattison, who was dean from 1990 to 1999, that increased resources needed for the recruitment of my replacement would not be available. He talked with Dr. Mattison about the possibility of relocating the faculty of this department to the school of medicine.

Our department had an external review the early part of 1992, and we received a glowing report. It recommended that the school (GSPH) ensure the continued survival and prosperity of a department that was essential to its mission. When the conversation between Drs. Detre and Mattison became known, a letter was written by the department heads of GSPH, addressed to Dr. Detre stating their objection to any dissolution of the department. On December 14, I received a memo from Dr. Detre in response to the memo of the department heads that he never considered relocation of the IDM faculty and that he would like a new committee to look at the situation of IDM as a function in the health center. This committee was chaired by Dick Simmons, and it included George Bernier, Ed Benz and Don Mattison (Simmons was chair of Surgery, Bernier was the dean of the School of Medicine, Benz was chair of Department of Medicine, Mattison was from the GSPH.).

In 1993 the appointment of a search committee for the chair of the Department of IDM was appointed. This blue ribbon committee, chaired by Herb Rosenkranz (chair, Department of Environmental and Occupational Health (EOH), GSPH), had on it Ed Wing, Ron Herberman (director, Pittsburgh Cancer Institute), and Dick Simmons. It had the approval of Dr. Detre but not his financial commitment.

This standoff in the search for the chair continued for two fruitless years. Candidates came and went, but no job commitment could be made because no additional funds were available. Finally on December 8, 1995, Mattison called in Herb Rosenkranz and myself and told us that he was giving up the idea of IDM as a separate department but wanted it incorporated as a "program" in Herb's department. Rosenkranz was chosen because he had been a card carrying microbiologist under Harry Ginsberg at Columbia University School of Medicine and was in charge of the microbiology course for medical students. I pointed out to Mattison that if finances were the only problem, he could maintain IDM as a separate department by appointing Chuck Rinaldo as chair. He is supported by the Department of Pathology anyway, and would not need additional financial commitments.

On December 20, Mattison asked me to call a meeting of the IDM faculty to tell them of his decision. At the meeting, instead of giving them a decision, he told them that they could either choose to remain as a separate department under Rinaldo, or choose to become a program in EOH. This statement was different from the decision that he had made when he informed Rosenkranz and me. Given this somewhat ambiguous message, it is not surprising that the faculty strongly opted for a separate department. Mattison left the meeting saying ambiguously that this too was his "preference".

Dr. Mattison did not follow up by talking with Rinaldo.

Instead he made the decision to dissolve the department on March 25, 1996.

This about face brought on the displeasure of the faculty. His waffling also weakened the strength of his own proposal. Bob Yee told him that "he would fight him all the way down the line". To carry out this decision, Don appointed a committee chaired by Ed Ricci, chair of the Department of Health Services Administration to submit a plan of execution, "due by fall". By inserting another committee under Ed Ricci in the course of events, Dr. Mattison created a new variable. After almost a year, on November 19, 1996, Ed came to report to our department. It appears that matters were complicated by the fact that the UPMC (medical enter) leadership (primarily Detre) separated the search for a new chair for IDM, from the recruitment of a director of a new entity, called the "Institute of Virology". Dr. Ricci thought IDM could be saved by somehow relating IDM to the "Institute of Virology".

At this point, I suddenly realized an obvious difficulty in the position of the leadership of the medical center. What they had been doing was to bypass the Department of IDM and GSPH while building a separate "Institute of Virology", whose main research assignment would be AIDS and HIV, when such a research commitment already existed in the Department of IDM, under the leadership of Chuck Rinaldo. Surely this could not be justified in GSPH or in the medical center or the university as a whole. Based on this enlightenment, I made the following remarks.

I began by pointing out the irony that the school of public health of the University of Pittsburgh is considering the elimination of its department of infectious diseases while other major schools of public health are talking about strengthening and expanding theirs. We have to recognize that these are times when infectious diseases are back in the mainstream of public health and medicine. I then noted the incredible fact that while the rest of the health center is talking about the establishment of an institute of virology of which IDM in some form will be a vital part, the Dean knows next to nothing about what is being proposed, and does not take part in such deliberations. Has he or the Ricci committee considered the likely possibility that the IDM group will be sequestered away from the school of public health, and put in a newly founded institute, with the school of public health giving up all the research overhead of the group but contributing all the space for its functions? It seems obvious to me that this school and specifically its Dean should be immersed in the negotiations about the establishment of the institute, and actually be a prime mover of its establishment if he wishes to keep for the school of public health a place in the teaching and research in infectious diseases.

I concluded my remarks by saying that the thinking and politics of the UPMC leadership should not play a disproportionately large role in his (Ricci's) thinking. He and his committee have an important role to play. They represented the school of public health, and have the job of expressing its opinion on this issue. To me the question they have to answer is: Does our school of public health need an IDM program, and if it does, what form should it take? After a year of deliberation this straightforward question has to be answered directly and forthrightly. To me if it could answer the question by saying that "the school of public health needs an IDM program, and it needs it in the form of a department". That would be about the best it could do and about all it could do. Dr. Ricci was nodding his head during the entire time I talked, and left me the impression that he agreed with me. In fact, he said several times, "I hear you". The other members of the department who were present were happy I made my remarks.

On November 21, there was a "cabinet" meeting (meeting of department heads, senior faculty, and Dean, GSPH) during which Ed Ricci reported what he and his committee had done about IDM. After Dr. Ricci's report, I made remarks similar to those I made at the departmental meeting. However, this time I pointed to the Dean's personal responsibility to represent at the higher levels his view and the view of GSPH about the need and necessity for IDM. Whether IDM remains a department should be and is his primary responsibility irrespective of what his superiors thought. Now that the element of an "institute" has been injected into the equation, this represented an opportunity and possibility to solve our problems of resources and small critical mass. However, I pointed out that in any formulation, IDM should be the main component of the institute, and hence he and GSPH should be in the front lines helping to formulate the shape and substance of institute instead of leaving its planning to others. I think it is fair to say that the audience was impressed by what I said.

Just before the Christmas recess on December 20, 1996, Dean Mattison came to my newly relocated office of "retirement" to tell me that he believed that the department will remain intact. Although there were still uncertainties, he did state this belief. Apparently, he was impressed by the strong resolution of the Faculty Senate that the department should be preserved. This stand was in turn partly propelled by an "impassioned" speech by me at a "cabinet" meeting, deploring the uncertain status of our department when infectious diseases are recognized nationwide and world wide as being increasingly important to public health and medicine. I would have been ecstatic with his statement were it not for the fact almost exactly a year ago we were in the same situation. That is he had then come to my office telling me that he had decided to name Chuck Rinaldo as the new chair of our department, and that the period of uncertainty concerning the department, then of about three years' duration, was over. Now after a whole year, during which he revoked the decision he announced to me, and waffled in indecision, he arrived at the precise same point.

Finally, the decision was made by the Dean! On February 7, 1997, he sent by e-mail a notice to the GSPH faculty and students that he has appointed Chuck Rinaldo as "chair" of the Department of Infectious Diseases and Microbiology. It took four years for this school to find my successor, and one year of agonizing over its survival. But I could not be happier about the outcome.

I began to understand the dynamics of the situation. Dr. Mattison's decision to eliminate the department could only be reversed by patience and persistence. It was necessary to mobilize the public opinion of the school. This took time. But in the meantime the department did not fold, as it could have done. The faculty of the school perceived the importance of infectious diseases and microbiology in public health, and insisted on the survival of the department. Whatever I might have done was my last contribution to this department and this school, and it was made after my official retirement when I was already off the pay roll.

After the departure of Dean Mattison, and the retirement of Vice Chancellor Detre and me, this Department has continued to survive and is flourishing under the chairmanship of Chuck Rinaldo, and deanship of Bernard Goldstein.

In the Spring issue, 2004, of "Public Health", the official publication of GSPH, the front cover showed a picture of four investigators of the Department of IDM under the title: "Four Pronged Attack: GSPH Researchers developing Novel Approaches for AIDS Vaccine". The investigators were Chuck Rinaldo, Phalguni Gupta, Simon Barrett-Boyd and Velpand Ayyavoo. The last two are new faculty members recruited since my retirement.

chapter

| 12 |

AN ASSESSMENT OF MY CAREER

M y actual retirement from work did not come until February, 2002, when I returned from what may be considered a second career in Taiwan (1997–2002, see Part III). But most of my work was earlier at the University of Pittsburgh. I retired from being Chief of the Division of Infectious Diseases in the Department of Medicine in 1992, from being Director of the Microbiological Laboratories at the Presbyterian University Hospital and the Department of Pathology in 1995, and from being Chair of the Department of Infectious Diseases and Microbiology at the Graduate School of Public Health in 1997.

By 1995, it was clear that my tenure at the University of Pittsburgh was approaching its conclusion. My colleagues at the Graduate School of Public Health and the School of Medicine decided to give me a gift. On June 9, 1995, they arranged a one day retirement symposium in honor of me called "Current Topics in Medical Virology and Antiviral Therapy". In the evening, there was a gala dinner of 190 participants in the Carnegie Museum of Natural History in Pittsburgh. We had dinner with orchestrated classical music under the skeletons of gigantic dinosaurs. During the dinner, a book of letters was presented to me from my old friends, colleagues at the university, and past and present students and fellows. No doubt, these letters emphasize my strengths and merits and they will not be quoted. Still, individually and as a collection, they portrayed some aspects of myself. This festive event and the contents of the letter were communicated to my son, John, who could not come to the festivities. He wrote me a candid letter giving me his assessment of my career. After presenting his views, I conclude with my own assessment.

Assessment by John Ho

John was then a 39-year-old associate of McKinsey Consulting Firm (see Chapter 6). I wrote him a detailed letter describing the festive events and quotes from the letters I received. He replied in a long, thoughtful letter. He expressed some candid opinions about my life and career from his point of view, and posed some pointed questions, some for the first time in my life. I provide excerpts from his letter and then my response.

Dear Dad,   August 27, 1995

About your letter, it seems to be a pretty good blow by blow account of the evening's festivities with details that I didn't know. As perspective, while we were growing up, we understood extremely little about your career or scientific interest, your progress and impact, your colleagues, and day-to-day activities. It was just a black box. You may have viewed it as uninteresting by us, or maybe in your mind, we couldn't really understand any of it.

When you did discuss it, it usually came with a ready assessment and a 45-minute discussion — great for post-docs, but smaller doses more often might have clued us in better. What I really missed understanding was the following:

- How your career progressed from post-doc/fellow through the professional ranks. How long did it take? How was it decided? What activities pushed you to the next level? (MH: See Chapter 10.)
- How did your responsibilities change as researcher, clinician and teacher? What sparked your changing research interests? Why?
- How were you perceived in the narrow virology, broader microbiology/ID and broadest medical research areas? Who were your contemporaries?
- What kind of a person really were you at work? What kind of researcher were you? What did you like, writing, doing the project plan, actually doing the experiments, analyzing data?
- How did students perceive you and how did you impact them? Was coming out of your lab a great resume builder?
- Lastly, how did you resolve your chosen career with your other interests and obligations, particularly the family? Did you feel like you had the proper balance? (I perceive you as a workaholic now and when I was growing up — many accuse me of this too.)
  Coming back to your letter, it gives me clues and hints in several of these areas. But like all good answers it only raises more questions.

- I seem to understand very broadly how your interests changed from cell culture/viral culture-interferon to CMV/EBV and HIV. I'm still unsure of the timescales and rationale for this switch nor if there was a lot more changes/interests along the way. Also, I don't get a sense of how your career path changed with it.
- I also seem to sense an extremely high admiration by your colleagues and students. It seems you were a good teacher, inspirer and mentor. Being in your lab seemed to propel many individuals' careers and gel their scientific interests. Again, did you perceive yourself to be a strong mentor or actively build your mentoring skills along the way or was it a "natural phenomena"?
- Also, it seems you contributed a lot scientifically, yet you discount it. Why? Science is like evolution, steady progression than a paradigm shift. Shifts are infrequent, one should not chide oneself for not coming up with them. Advancing the ball other ways can be as impactful. I wonder what John Armstrong or Bob Yee (colleagues at the Graduate School of Public Health — GSPH) would say of their contributions. (MH: It is true that at the banquet, I gave a speech essentially mitigating my scientific accomplishments.)
- It seems many had professional recollections of you yet personal insights are few. This seems to mean a high degree of professional admiration (as this is the first thing that comes to mind for them and everybody had something good to say). It in addition suggests a high degree of work-orientation and emphasis in your life which as I mentioned before is good or bad, depending on how you feel about it.
- Also there seemed to be few non-professional acquaintances there. How do you feel about this?

On specific things that you raised:

- The outside speaker series seemed too great. The contrast between yourself and Tom Merigan was interesting — him into the mafia, you "in the wilderness." (MH: This comment comes from my description. See also Chapter 8. Tom was unquestionably a leader in our midst.)
- The reception after the seminar seemed particularly well attended with lots of people (few of whom I know).
- The dinner was clearly the highlight and I will always regret not being able to attend. McKinsey is a cruel mistress which makes you miss all sorts of personal milestones. I wonder if it's worth it.
- The initial speeches seem very thoughtful. One that particularly sparked my interest was your picking Bill Pasculle to head the labs.

(MH: See Chapter 10.) How did you decide, what criteria did you use, how did you "size up the man"?

■ I wish I had seen Bosko. (MH: Postic, my doctoral student, later Professor of Medicine in charge of infectious diseases at the VA Hospital, University of Pittsburgh.) Haven't seen him in years and I know that he (along with Dr. Armstrong) were your longest associates.

■ Charles Chu seems to be the unique "friend only" speaker. You two have had a unique relationship. I believe you are more personable, relaxed and open when you are with him than anybody else (including your family). He really is your brother (do you ever wish you had one?). (MH: More about Charles and his wife Bettie in Chapter 19.)

■ Bob Yee and Bob Atchison (both faculty colleagues in our Department of Infectious Diseases and Microbiology at GSPH) are two I also wish I had seen... both were pivotal in my life — I probably would be flipping burgers in McKeesport without them taking an interest in me. Thanks for putting them on the case, they are two of the greatest people. (MH: These two helped John perfect his science exhibit, "Transfer of Genes in Bacteria," which won the Science Fair Grand Prize in Pittsburgh in 1975.)

■ Your speech seems typical for you, very philosophical and self-critical. I always wonder what you really feel about yourself, i.e., your ego structure — are you an insecure over achiever? As I remarked earlier, you really give yourself a hard time on your contribution level, unwarranted in my mind.

■ Also you raise some issues about prejudice against Asians — specifically Chinese — had it affected you personally and made you take certain directions? Is it why you entered medicine, where relationships/networking are less important to success while true merit is? I've always thought so (also it fits your need to do something noble and worthy such as good service).

■ The comments in your memory book (MH: collection of letters) I found interesting. My thoughts:

- You have always seemed to be one of the "old breed" of renaissance/Confucian scholars, tremendous breadth of knowledge particularly in philosophy, religion, analytic thought (logic), history and current events. I have met only a few like you.

- In some ways I took a more practical bent to my training — engineering, science, practical skills like cooking, enology — because I felt I could never compete with you in the scholar/philosopher arena. I feel comfortable in it but marvel at your grasp of languages, history and philosophy. Perhaps Gregory (MH: his son) can be a blend of both — let's hope so.

- Again mentoring seems to be a theme repeated. Also your rigor and laser-like ability to dissect a research project. You must have helped a lot of people become better scientists — they also probably were very nervous presenting things to you.
- Later comments seem to me to be a classic illustration of how medical/biologic research has changed. From phenomeno-logic/statistical research to genomic mapping, signal trans-duction assessment and receptor theory. Those you have trained will be the ones struggling to bridge the traditions. It will be like me and Gregory, him 1,000 times more com-puter/multimedia literate than I.
- Do your Chinese students hold you in different regard?

Just a few other questions:

■ Do you ever wonder what your life would be like if you had returned to China? (MH: This question is answered in Part III.)
■ What would you have done differently given another chance?
■ What do you want to do going forward?
■ What does life mean to you? I haven't figured this one out yet but I've kind of entered my mid-life crisis early — helping Fortune 500 companies get richer at McKinsey doesn't seem to be a life ful-filling work.

Love & regards,
John

My Assessment

Dear John,   September 1, 1995

Thank you very much for your letter of August 27. I was beginning to think that you would have no response to the report I gave you. But was I ever wrong! I read through your letter many times with intense interest. Some of your questions are very penetrating; easy to answer superficially but quite difficult to answer comprehensively. One of my thoughts was that nobody was really interested in these things about me. But your ques-tions have emboldened me to sit down to think about them, and to answer them as best as I can. In answering them, I believe I can make a contribu-tion to the study of problems of personal and professional life... I feel that you are at a stage in your career when you wish to understand how I might have reacted or behaved at your stage... In this letter, I will only touch on some of the highlights of my answers. I will proceed in the sequence of your remarks in your letter.

It is clear that you are in the midst of the "Sturm and Drang" of partnership elections. Election would be achieving something you want, but more to the point, it would signify success and accomplishment in your world. On a larger scale, you wonder (later on in the letter) whether work at McKinsey is your "life-fulfilling work". All this is natural. Even your doubts are healthy. The world of your work should not encompass all your ideals and aspirations. In this light, a successful election as a partner may not be all that important. You have adequately demonstrated, irrespective of the outcome, what you can do and accomplish in the business world. Look at the outcome of the elections as an aid to your future career choice rather than a sign of success or failure. . .

You asked me what determined my career progression, choices and advancement. I can answer that with one word, "orthodoxy". I have usually chosen the accepted, least risky, not necessarily always the easiest, route. As an illustration, I have trouble understanding people who seem to flaunt orthodoxy and conventionality. Of people I really care for, I cannot understand the behavior of Kevie Chu (MH: son of Charles Chu, life time friend). Why does a man of his talent make so many unpredictable and apparently senseless twists and turns in his career? I consider this lack of understanding exactly that, i.e., a defect on my part rather than his. For Kevie does well for himself, he seems to have had a consistently happy and fulfilling life. By that test, he is at least as much a success, if not more so, than I am.

I did all that was demanded of me, by society, by culture; implicitly by my father, to be a success, academically and in my career. Underneath, my real self may have been at times buried. It is important to understand and look for my real self. Where has it been during this drab and relentless pursuit dictated by conventionality? I can only touch on some aspects of my probable real self. I remember being inspired by the study of philosophy at Harvard. But I gave that up to pursue a safer (even though a meritorious) career in medicine. I remember having a number of moments of scientific inspiration and ecstasy in research during my fellowship. But I allowed myself to be dissuaded by Enders to go on to Rockefeller University to get a Ph.D., which would have prepared me better for basic research. But I launched myself instead on a conventional academic medical career path at the University of Pittsburgh.

Promotion and advancement in such a career requires excellence in teaching and research. I remember moments of inspiration when I lectured. I could provide sparks if I wanted. At times I thought I had real talent for understanding and managing people. Administration seemed to come easy to me. As in the case of scientific mentoring, I did have a knack for perceiving the essence of a problem. I felt inspired when I was able to do this. I believe these moments of inspiration and elan, of which I had some, but not many, hold the keys to my true self.

My ambivalence toward research is deep rooted and confusing. By conventional criteria, I have been a success. That is measured by my rapid promotions, election to the right societies, many papers, invitations for lectureships, and continuous funding by the NIH. But I have not been able to entirely keep up, let alone lead, in modern biomedical research. In my pursuit of trying to be the conventional "triple threat", the prototype of the old fashioned medical academic, I have fallen behind in each of the "threats", whether teaching, research or medical practice. Realizing this produced boredom and ennui in the course of pursuing each of these activities. Why should this bother me? Well, one reason is, it's the same old thing, isn't it? Lack of conventional accolades of success. It's hard to get out of the circle.

> That is all for now. With love to you, Michelle and Gregory,
> Bätch (Dad)

The above letter, written in 1995, shortly before my retirement from the University of Pittsburgh, attempts to answer some key questions about myself. Have I expressed my real self in my life-time? Or has my education and career entirely been legislated by what was conventional? I seem to have said that both statements are true. I admit I was overly harsh in my blanket generalizations about myself during my speech at the banquet. Still, it is an expression of my mood then. I was also truly humbled by the event.

My ostensible reason for choosing a career in medicine was stated in Chapter 5. Of all careers, it is safest in providing the maximum of good and a minimum of harm. But John was perceptive in asking whether I might have chosen it because in medicine, "networking" is less important and "true merit is". Asians in the United States, specifically the Chinese, have been accused of choosing careers which demand a high degree of technical know how, and a great amount of education, but they have avoided the arts, politics, and the business world. There may be an element of truth in what he says, although I was not aware of this when I made my choice.

John's letter brings up the question, what were my real interests and skills? I believe that the incident at Brooklyn Tech I recounted in Chapter 1 is an indication of my true self. When I was 14 years old, I was mediocre in "shop", but I excelled in discursive writing.

I was punished by Mr. Kirkwood, our teacher of shop, who assigned me to write an essay on, "why I should not play at shop". Mr. Kirkwood gave me a "C" in shop but requested the English teacher to give me an "A" in English.

As I pointed out in my response, I was giving only "highlights" of my own assessment. But after giving the matter further thought, I feel I did make a few decisions regarding my education which were risky and unconventional, and to that extent, they might be expressions of my real self. I made the decision to return to war torn China in 1941. My ostensible reason was that I wished to remain Chinese. I still feel it was the true reason. In college, I made the unconventional decision to switch majors; from prestigious chemistry to what was outwardly less well regarded political science at Tsinghua University. People may suspect I was trying to follow my father's footsteps. It is possible that that was part of my unconscious motivation. But my ostensible reason as described in Chapter 3 still stands. It was to solve a nagging intellectual question, that is, what was the best political system for China? This may smack of wild eyed idealism nowadays, but my subsequent two years at Harvard College prove that it was a true quest. There I shifted to major in an even less practical field, philosophy, far away from Father's pursuits. The denouement of my interest in political theory was my senior honors thesis. That was a response, even if not a perfect answer, to my nagging intellectual question. The thesis comparing two different perceptions of social sciences was my first success in academic research, and it offered me a sense of satisfaction that has remained with me throughout my life.

My foray into the social sciences in college was also an expression of a real interest and of my real self. My modest achievement in this area may be traced to my ability to think and solve problems analytically and to express my analyses in my writing. This is not inconsistent with what was discovered about me when I was 14 years old. Despite my having changed to medicine later, these same skills have fostered whatever I might have achieved and written as a medical scientist.

Switching to medicine might have been a radical change, but later I realized that medicine is such a broad field that it can encompass any number of diverse abilities and interests. I was not attracted to subspecialties requiring mechanical and manipulative skills such as surgery, and I eschewed those without further ado. In my fourth year at Harvard Medical School, I was turned on by case studies in neurology, psychiatry, and metabolic problems. I was interested in problem solving, in workings of the mind, and in research. Choosing internal medicine was natural. Entering academic research medicine was also an easy choice. I believe these choices were compatible with my true self as expressed from age 14.

Besides native ability and interest, another factor in making decisions concerning one's education and career is one's motivation. As I have mentioned in Chapters 1 to 3, a primary "Leitmotiv" in my life, developed in my youth and preserved throughout, was my desire to be of service to China. This aspect of my self was apparent from my decision to return to China for my education at age 14, and my shift of majors in college to political philosophy. However, after choosing medicine as my career, I had no opportunity to serve the Chinese people for fifty years. But the desire and hope to do this never left me. The nature of medicine as a career is such that its service orientation can be easily redirected.

After my retirement from the University of Pittsburgh in 1997, I had the opportunity to go to Taiwan to do medical research and to carry out a program of medical problem solving. This project, described in Part III, was a project that satisfied my life long interest to serve the Chinese people utilizing my professional ability. That is a type of fulfillment in life that is often hoped for but not often achieved. I also got the opportunity to learn and to teach. I had a refresher course in Chinese culture. I explained and expounded my thoughts among new friends and in the lay press. To be able to do this was made possible by a life long interest in humanistic studies. The impact of my work could not have been achieved in five years in the United States. Taiwan's smaller size and smaller community worked to my advantage. Thus late in life, I achieved something that gave me complete satisfaction in self expression. I am deeply thankful for having had that opportunity.

part

III

EXPERIENCE IN TAIWAN

INTRODUCTION

E ven in subtropical Taipei, it was cold and brisk on January 17, 2002. That night, President Cheng-wen Wu, of the National Health Research Institutes (NHRI), who had become a close friend, gave me a farewell dinner at the American Club. Besides Carol and me, he invited about twenty of the senior staff of NHRI and their spouses. I was retiring for a second time, and returning to the United States after working in Taiwan for almost five years. The occasion was nostalgic and intimate.

Actually, this represented my farewell to my entire "Taiwan experience". When did my Taiwan experience begin? This is a complex question, particularly in these days of shifting international relations. At this point of my writing, Taiwan may be on its way to becoming an independent state. My experience in Taiwan is connected with Taiwan's Chinese heritage. This heritage is a fact that cannot be changed by Taiwan's political future. Taiwan's people, Taiwan's language, Taiwan's institutions, Taiwan's culture are all Chinese; not the China of the Ching Dynasty, or the China of the Kuomintang Nationalists, or the China of Communist China, but Chinese. Should Taiwan become independent, it would be the second Chinese state. I was not born or educated in Taiwan. I have no roots in Taiwan. Yet I feel attached to Taiwan.

Why is that? One can point to my upbringing, which was oriented toward being Chinese. One can point to my father, who worked for the Chinese Nationalist Government in Taiwan from 1948 to 1972, whom I first visited in Taipei in 1965. My real connections with Taiwan were developed after I was elected to Taiwan's Academia Sinica in 1978. They were solidified by my five years of work at NHRI.

During this period, I directed and organized the Division of Clinical Research of NHRI. Its primary purpose was research of the antibiotic resistance problem, and advocacy of a national policy to alleviate it. I chose this problem as my project because it was a serious problem, but it was also one I might do something about because of my training. Besides my primary mission, I became inevitably involved in public health problems in infectious diseases and in medical education. In 1998, I recognized Enterovirus type 71 as the primary culprit of a major outbreak of infantile sudden death and hand-foot-and-mouth disease in Taiwan. At the end of my tenure, I received the first prize for research excellence at NHRI, with an award of NT 100,000. I also received a medal of achievement in public health, first grade, from the Taiwan Department of Health.

Tonight, the ambience at the party would be intimate and nostalgic. For these five years, I had gotten to know my friends at NHRI well. The American Club is an uncommon site for such occasions. They have to be sponsored by a member. It is in a high class area in the suburbs on Taipei, near the famous old five star hotel, the Grand Hotel, previously owned by Madame Chiang Kai-Shek. Ordinarily, such occasions are held in a Chinese restaurant in town, where an over-generous dinner of ten or twelve Chinese courses is served. In this case, there was a cocktail hour before a sumptuous sit-down dinner of American food, held in two adjoining rooms dedicated to the occasion. Vases and baskets of flowers adorned the floor and tables. After dinner, almost every one of the ten or so principals spoke.

President Wu began by recounting how I began my work in Taiwan. I began coming in 1980, after I was elected to Academia Sinica when I was on the preparatory committee of a new institute of the Academy, the Institute for Biomedical Sciences (IBMS). I began to come yearly to Taiwan after 1992, to direct a post-doctoral program for infectious disease fellows. I became well acquainted with Taiwan's health care program and its problems. The NHRI, patterned after the NIH in the United States, was established in 1996, and he wanted me to come to direct a division in July, 1997, after I had retired from being chair of the Department of Infectious Diseases and Microbiology in the Graduate School of Public Health, and the chief of the Division of Infectious Diseases in the School of Medicine at the University of Pittsburgh. I accepted, after presenting a specific mission and a plan of execution. I came and carried out my mission. My division had become a model for NHRI; because of its clear, steadfast mission and purpose.

Cheng-wen then thanked me because I had also changed his life. He was a medical graduate from the University of Taiwan, and a Ph.D. in biochemistry from Western Reserve University. By 1988, he had become a chaired professor of biochemistry at New York State University at Stony

Brook. That year, he was nominated to be the first director of the Institute of Biomedical Sciences (IBMS). That is when he returned to Taiwan permanently. This came about partly due to my doing.

When my turn came to speak, I recounted for the first time in public how his recruitment came about. In 1981, a group of five Academy members in the United States, Drs. Paul Tso, Shu Chien, Shih-Hsun Ngai and myself, worked together under the leadership of Paul Yu, the well known chief of cardiology at the University of Rochester and the erstwhile personal physician of Chiang Kai-Shek, to plan for and build a new research institute, the Institute of Biomedical Sciences (IBMS), for Academia Sinica. The work progressed rapidly, and by 1986, a permanent staff and a new six story building were in place. The time had come to nominate a permanent director. There were two candidates. The first was Paul Yu, who had nominated himself. His background and achievements were impeccable. Without him and his connections at the top, IBMS would not have been established. Unfortunately, it was apparent to all of us who worked closely with him that he had deteriorated mentally in the last few years, and he had also been diagnosed as having Parkinsonism. The other candidate was Cheng-wen, nominated by Shu Chien and supported by the others except Paul Yu. Cheng-wen was much junior to Paul in stature, but he was young, vigorous, eminently qualified and he had proven himself as temporary director. The problem was how to manage this election without offending Paul Yu, who was a member of the voting group. This is especially delicate among Chinese, where seniority both in age and in qualifications is not only highly respected but preferred. I was chosen to chair the nominating committee, which consisted of the five aforementioned persons. Since some of us were in Taiwan, and I, among others, was in the States, we chose to meet by phone. I opened the meeting first by asking for the names of the nominees. Paul Yu and Cheng-wen Wu were named. As chair I could choose which candidate to vote on first. I chose Chen-wen hoping to settle the issue without having to vote on Paul. I decided to ask each person individually whether he was for Chen-wen. I began with Paul Tso, and then went on to Shu Chien and Dr. Ngai. Each one of them cast his vote for Chen-wen. Without expressing my own vote, Cheng-wen already had a majority of the votes. Then I turned to Paul and asked for his vote. At that point, he tactfully ignored his own candidacy and also voted for Cheng-wen. So the vote for Cheng-wen was unanimous, and there was no need to take a vote on Paul Yu. Hard feelings were avoided, and Cheng-wen was nominated. This choice was subsequently endorsed by the President of Academia Sinica.

I then told of my own aspirations in Taiwan. Although I had been in Taiwan less than five years, they represented fulfillment of a lifelong ambition, an opportunity to work for the Chinese people. While most Chinese are on the mainland, Taiwan represented China to me. I was given a unique opportunity to work and serve in Taiwan after I was elected to Academia Sinica. Academicians are expected to help Taiwan. This "obligation" became my welcomed opportunity. My training in medicine and my career in infectious diseases uniquely qualified me to serve, as my work in the last five years showed. Service was my original objective when I chose medicine as a career, which I chose because it was one career I could think of without the fear of doing harm.

chapter

| 13 |

ACADEMIA SINICA

One day in September, 1976, after termination of the biennial meeting of Academy Sincia in July, I received in Pittsburgh a long-distance telephone call from Dr. Paul Tso in Baltimore. Largely because of my research accomplishments in virological research, especially in interferon, I had been one of the candidates considered for election. I had heard that my election to the Academy had failed. I was deeply disappointed, but I was not expecting a call from Dr. Tso, a member of the Academy whom I did not know.

"You were a candidate who got the largest number of straw votes at the biennial meeting of the Academy in Taipei in July. When it became apparent during discussion of your candidacy that you probably would be elected, one of the very few members of the Academy who knew you personally, C.C. Li, got up at the conference and questioned your 'maturity.' That was the word in English he used. At a result, your election failed."

This mysterious innuendo coming at that time in Taiwan could only represent a questioning of my political stance. Taiwan was then ruled by the government of Chiang Kai-shek, who had fled communist mainland China. There was paranoia against leftist and even liberal political views, immeasurably worse than in the America of the McCarthy era. These views were suspect because of the fear that they may be related to communism.

My father was then still working for the Nationalist Chinese government in Taiwan. I myself had been a student on the mainland, and like many students, I was critical of the Nationalist government. The criticism was primarily directed at the incompetence and corruption of the Chiang government rather than its anti-communism. While it is true that I often voiced my dissatisfaction among friends in private, I was basically a scientist and had never participated in any type of political activity. Most of the members of the Academy did not know me personally, and they had no way to ascertain the facts. They did not give me the benefit of the doubt, and voted against me.

"However, in view of the excellence of your scientific work, a group of us would like to help you overcome this injustice. Dr Yuan I-ching and I are willing to fight for your election during the next meeting of the Academy in 1978. But first I would like to ask you whether you are willing to accept the responsibilities of being an elected member. Academia Sinica is different from other honorific societies, where no responsibility is involved. You have to promise to work for the scientific development of Nationalist China (Taiwan)."

I said, "I will gladly accept this responsibility. In fact, it is welcomed because I have always wanted to have an opportunity to serve the Chinese people."

After this call, Dr. Tso and Dr. Yuan thought of a very clever scheme. They wrote a letter to the Nationalist Chinese embassy in Washington, D.C., asking for information on Monto Ho; in particular, whether he had participated in any subversive political activity against Nationalist China in the United States.

Dr. Yuan I-Chin was a veteran member, who was elected in 1948. He was a graduate of PUMC (Peking Union Medical College), who later obtained a doctoral degree at Johns Hopkins University in public health and biostatistics. He became a well known figure in public health in China. Dr. Paul Tso was younger and had been elected in 1972. He was a professor at the Johns Hopkins University School of Public Health and a well known biophysicist.

It is well known that it is difficult to prove a negative statement. It is doubly difficult to prove that somebody was not involved in some subversive political activity.

As one can imagine, when the bureaucracy of the embassy receives such a blanket request, they would probably leave it unanswered unless there was clear cut evidence that the person in question was truly subversive. And that was what happened. Drs. Tso and Yuan utilized this lack of response to my advantage by announcing during the meeting in 1978 that they had inquired at the Embassy whether I had a record of subversive activity, and discovered that I had none. Therefore, they concluded, there was no truth to the accusation that I was pro-Communist. In this way, the accusation against me was dissolved, and as a result, I was elected in 1978 by a large majority.

In 1980, I attended my first meeting of Academia Sinica in Taipei. This was the first time my wife and I attended what was for us a national Chinese gala affair. Taiwan was developing at the time economically, but basically people were still poor. The amenities offered to us were truly surprising. Each couple received first-class airplane tickets, and, while in Taiwan for an entire week, a personal, private car with a full-time driver. We were put up in the luxurious Grand Hotel. We felt exceedingly honored by the then-President Chien Shih-liang. The succeeding presidents of Academia Sinica, Wu Tayou and Lee Yuan-tseh, have decreased the amount of material amenities, but the honor shown members of the Academy during each biennial meeting is still considerable. It is always a national event.

Academia Sinica was founded in Nationalist China on the mainland in 1928. It moved to Taiwan in 1948. Its original model was the French "Académie", which is a government supported elite institute of learning. Most countries have such organizations, such as the U.S. Academy of Sciences, the British Royal Society, and the Russian Academy of Sciences. The main difference among these institutes is that some of them have an actual research component. The U.S. Academy of Sciences does not. From the beginning, Academia Sinica has established research institutes encompassing different scientific disciplines. Originally twenty were envisaged. The number now has increased to twenty-four. There are now six institutes in the life sciences alone, three of them relating to medicine, the Institute of Biomedical Sciences, Institute of Molecular Biology, and the Institute of Biochemistry.

One function shared by academies of all nations is an honorific one. Membership is accorded to their own scholars and scientists of national repute. Every two years, during a meeting of all its members, Academia Sinica elects a maximum of ten members in each of three areas of learning, the physical and mathematical sciences, the humanities and social sciences, and the life sciences. This maximum quota is almost never attained. There are physicians and engineers among the members, but they are elected more for their achievements in research than for their professional achievements.

While Academia Sinica is a fairly recent organization, its roots are ancient in Chinese culture. Academic and intellectual achievement is highly prized among the Chinese, and it had been recognized by a government supported examination system that goes back two thousand years. Academia Sinica and its counterpart on the mainland (also called Academia Sinica) seem to be successors to the Imperial College of dynastic China, which consisted of scholars who had passed the imperial examinations at the highest level. These institutions bask in an extra measure of esteem.

Many of the members of the Academy at the time I was elected were born outside Taiwan, although most recently elected members had been educated in Taiwan. During the last twenty years, there has been a gradual shift of elected members. They were not only educated in Taiwan but were also born in Taiwan. This is a reflection of scientific maturity in Taiwan. I believe that members of the Academy, without regard to their origin, have worked diligently and without discrimination for the development of science in Taiwan.

The Establishment of IBMS and NHRI in Taiwan

During the biennial meeting of Academia Sinica in 1980, two important resolutions were passed. One was the establishment of the Institute of Biomedical Sciences (IBMS) and the other was the establishment of the Institute of Molecular Biology. Both were directed at promoting the "biomedical sciences", which was then a new and unfamiliar term in Taiwan. The successful establishment of these two institutes would mark the beginning of an upsurge of research in biomedical sciences and biotechnology in Taiwan, and they set the pace for the entire nation for the next twenty years.

After passage of the resolution in 1980, a group of five young Academy members in the United States, Drs. Paul Yu, Paul Tso, Shu Chien, Shih-Hsun Ngai and myself, formed a working team to plan for IBMS. We began to meet every 1 or 2 months in New York, at the Presbyterian Hospital of the Columbia College of Physicians and Surgeons.

"We have to lay out every aspect of the institute. Let's begin talking about its mission and purpose, all the time thinking about how to develop personnel policy and the physical plant," said Paul Yu at the first meeting.

Paul was a charming, handsome man of medium stature. He was a brilliant speaker, both in English and in Chinese. Though a native of Jiang Xi in central China, he spoke both English and mandarin without an accent. He was a graduate of Shanghai Medical College, one of the best national medical schools in China. After coming to the United States on the prestigious Boxer Indemnity Scholarship, he eventually rose to become head of cardiology at the University of Rochester Medical School. He was later elected president of the American Heart Association. Paul was a brilliant clinician besides being a good researcher. He was the personal physician of President Chiang Kai-shek and his son and successor, Chiang Chin-kuo. Because of his access to the younger Chiang, he had considerable clout politically. He was the one who was responsible for acquiring the required funds for IBMS from the government. Without him, we would have been bogged down by bureaucratic red tape which might have aborted our project altogether. Unfortunately, Paul died in 1991.

Our meeting place, the Presbyterian Hospital in uptown New York, was where Drs. Chien and Ngai had their offices as professors at Columbia College of Physicians and Surgeons. It was very convenient for me, as I was but an hour's flight away in Pittsburgh. We would start the meeting in the morning and finish in the early afternoon. Mrs. Ngai, who also worked at the hospital, took care of our lunch arrangements. We would either eat in the dining room, or she would have lunch brought in from the outside for us. The four of us were idealistic but realistic. We were very congenial, and felt we were striving for a common purpose. We were familiar with the problems in Taiwan, but we knew what the ideal situation was, in the United States, and we strove for it.

"The institute should be directed at specific diseases," said Shu Chien.

"We might use the U.S. National Institute of Health as our model", I said.

Shu Chien was the son of ex-President Chien of Academia Sinica, and like his father, he has the graciousness of a Confucian scholar. He is on the small side, but he is attractive and elegant in demeanor. He was a medical graduate of University of Taiwan, and became an eminent physiologist at the University of Columbia, specializing in rheology, or the study of blood flow. More recently, he has led the bioengineering group at the University of California, San Diego. Dr. Chien is efficient in everything he does, and he excels in both English and Chinese. In addition, he is versed in the use of the technical Chinese language (officialese) that is required in official communications with the government. We were able to do this rather than go through the bureaucracy of Academia Sinica.

Paul Tso said, "We are going to have trouble recruiting properly trained medical scientists. We have to start training them ourselves."

Paul's English and Mandarin has a Cantonese inflection. Paul is a native of Guangdong, and he had his earlier education in Hong Kong. He is also on the small but plump side, and exudes energy and forcefulness. Being a professor of biophysics at the Johns Hopkins University, he was the non-physician in our group. He introduced me to Academia Sinica. He was very imaginative and had many innovative ideas. He pushed for training of clinicians in the specialty of oncology, which has since become an established medical specialty in Taiwan.

By 1986, a six-story building was built according to our specifications. Dr. Ngai specialized in construction and equipment purchases. He was assisted by Chai Tso-yun in Taipei, also a member of the Academia of my generation. Ngai was Professor of Anesthesiology at Columbia. He died in 1999.

The concept for IBMS was based on the National Institutes of Health in the United States. The mission we initially envisaged was research in cancer, cardiovascular, and infectious diseases. Dr. Tso and Dr. Chien were responsible for the development of the first two areas, and I was responsible for infectious diseases. Drs. Chien and Tso put together the research programs in cardiovascular diseases and cancer in the early phases of IBMS. Dr. Chien has remained chair of the advisory committee of IBMS.

My contribution in the group was the schematic organization of IBMS. Our model was the NIH in the United States, although much smaller in scope. Since I am not primarily a molecular biologist, my personal contributions were in the general medical and biological aspects, especially the microbiological aspects of biomedical sciences. Initially, we concentrated on intramural research, but like at NIH, we were always aware that there would be an extramural research component. This was pursued at the same time as intramural research. Only quality and state-of-the-art research would be pursued. Since this was not always achievable, we realized the importance of training in the pursuit of excellence. Many projects in the future were done by people that we trained. Dr. Tso planned very early for training in oncology, and I envisioned the establishment of a training program in infectious diseases (more about this in Chapter 13 and 16).

As in the NIH in the United States, we planned for clinical investigation. This was difficult logistically, because Academia Sinica is in the eastern suburbs of Taipei, far away from any large teaching hospital. Initially, we thought of establishing a small research hospital attached to IBMS. Eventually, we solved the problem by contracting clinical research in the three designated mission areas — cancer, cardiovascular diseases, and infectious diseases — to three major teaching hospitals in Taipei; that is, the University of Taiwan Medical School, the Tri Services Hospital of the National Defense Medical College, and the Veterans General Hospital. This continued for about ten years, until IBMS was fully staffed.

By 1986, IBMS was a going concern, even though most of the professional staff were temporary and on loan from various institutions in the United States. Dr. Wu Cheng-wen was the last temporary director on loan from the New York University at Stony Brook. He decided in 1988 to cut his ties with the United States and to become the first permanent director of IBMS. This momentous personal decision had great impact, not only in Academia Sinica, but in Taiwan as a whole. It showed that an accomplished Chinese-American scientist, established in the United States, was willing to return permanently to Taiwan. This set a precedent and many others followed suit, not only in Academia Sinica, but also in Taiwan's universities and commercial companies. By 2000, IBMS had become an institute of 453 employees, with 217 professional scientific personnel, including forty-three investigators with doctoral degrees. In ten short years, IBMS became the largest institute on the campus of Academy Sinica, where there are twenty-three other research institutes.

The passage of resolutions during Academia Sinica's biennial meetings does not always mean that they will be carried out. Very often, members will pass resolutions that are unrealistic or represent wishful thinking. They often leave to the staff of the Academy the job of carrying out the resolutions as they see fit.

The difference in 1980 was that the members who proposed the two resolutions establishing IBMS and the Institute of Molecular Biology (IMB), mostly recently elected Academy members, including me, were bent on carrying out these resolutions ourselves.

If we look at the members of Academia Sinica elected during the ten years around the time of my election, a majority were Chinese-Americans who had achieved academically in the United States. They were people of my generation, born in China or Taiwan who had completed their graduate education in the United States. Like me, many were eager to serve the Chinese people and looked for opportunities to do so. This common interest made Academia Sinica more than an honorific society. It became for a group of us a service fraternity. That group found its focus of interest at my first meeting in the Academy.

A common problem, encountered by both newly established institutes, was the lack of senior scientific research personnel and leaders to direct them. In the beginning, the directors of the two institutes were assumed by young Academy members elected from 1972 to 1986, who were already settled in the United States. They were assigned on a yearly or even monthly basis. The Institute of Molecular Biology was successively led by Drs. Paul Tso, Chien Ho, Ray Wu, James Wang, Ru-chih Huang, and Dr. Wang Ching-chung. IBMS was led by Drs. Paul Yu, Shu Chien, and Dr. Wu Cheng-wen. Each one of these individuals was a prestigious professor in an American university. To recruit a research staff, we created a network among the

Chinese American scientific community in the United States and disseminated information about the two institutions. Researchers were recruited to go to Taiwan to do research, even if only on a temporary basis.

The artificial implantation of institute leaders from the United States lasted about ten years (1980–1990). Simultaneously, the biomedical sciences were developed in many universities in Taiwan. The common methodology of research was usually molecular biology. By the 1990s, molecular biology had become firmly established in Taiwan's colleges and universities, to the extent that it became the vanguard of all biological sciences. Taiwan was able, despite its small size, to participate significantly in the global genome project of the 1990s. It is now able to compete internationally in biotechnology, after having established competence in the electronics industry earlier.

After planning for IBMS for about two years, it became apparent to us that IBMS could not entirely emulate NIH. Being remote from any teaching hospital and located within the basic science-oriented campus of Academia Sinica, it could not avoid becoming an institute largely of basic science, inadequately related to clinical medicine. Thus began the job of planning for a separate institute, outside of Academia Sinica, devoted to the more applied medical sciences, such as environmental health and health policy. That was to be the National Health Research Institutes (NHRI), eventually established in 1996. The initial blueprint for this organization was written by me when we were planning for IBMS under Paul Yu in the early eighties. It was necessary to build a separate institute for applied medical sciences, in order to simulate the full scope of NIH. The actual planning for NHRI was accomplished in the late 1980s by Dr. Wu Cheng-wen, who became its prime mover. He was guided by an advisory committee led by Dr. Shu Chien, of which I was a member. In order to allow for flexibility in personnel policy, such as salary ranges, NHRI was funded as a foundation outside of the government. It was officially established in 1996, with Dr. Wu as president. Our infectious disease training program, that began in 1992 under IBMS, was transferred to this institute. NHRI has ten intramural research sections: oncology, genomic medicine, environmental and occupational health, geriatrics, health policy research, biotechnology and drug studies, vaccine studies and infectious diseases (clinical research). NHRI also took over the highly developed extramural research program and peer review system developed under IBMS. It is similar to the NIH extramural research program.

chapter
| 14 |

WHAT'S IN A NAME?

It was around 10 p.m. in May, 1998, when Carol and I were visiting with our son John, Michelle, his wife, and our five year old grandson Gregory, in their house in Wellesley, Massachusetts, when I received a telephone call from NHRI in Taipei. That's the usual time for them to call, because of the twelve hour time difference. There was an epidemic of hand, foot and mouth disease among children in Taiwan. Two laboratories in Taiwan had isolated a virus from the throat swab of two children with hand, foot and mouth disease. They successfully cultivated this virus in cell cultures, and identified it by using specific monoclonal antiserum against Enterovirus 71. The media people were asking questions of NHRI, which was the prestigious agency with experts in infectious diseases and virology. What they wanted to know was, in view of a disagreement as to the significance of the isolations, what was the meaning and significance of this virus isolation. I was asked to draft such a statement for NHRI.

The beginning of the major outbreak of hand, foot and mouth disease was in April, 1998. By the end of the summer, there were 129,106 cases reported throughout the entire island, and 78 young children would die. This reported number was estimated to be only ten percent of the actual number of cases.

Emergency rooms and outpatient departments of major hospitals throughout Taiwan were jammed day and night with mothers and fathers bringing in their children with hand, foot and mouth disease. This is a febrile skin eruption consisting of red blotches and small blisters on the hands, feet and mouth of a patient, usually a young child. In medicine, this clinical picture with simultaneous eruptions in these three locations is virtually diagnostic of an illness, due to a group of viruses called "enteroviruses", or viruses that multiply in the "enteron", or the intestines. Most of the children were discharged after receiving symptomatic treatment, as there is no antiviral drug against enteroviruses. A few patients with high fevers were admitted to the hospital, and observed for the development of complications.

And there were complications. What was unusual about this outbreak of hand, foot and mouth disease was that there were deaths, and unusual deaths. Some children began to cough up blood. They became short of breath, bringing up bloody, frothy sputum. They were bleeding from the lungs; they had pulmonary hemorrhage and pulmonary edema. Although they were rushed to the I.C.U. and received oxygen and assisted respiration, they died suddenly within hours of admission to the hospital. Heroic measures, such as the administration of gamma globulin, hoping that there might be antibodies against this virus in the pooled commercial product that might kill the virus, were of no avail. No one had heard of this type of sudden death due to an enterovirus infection. Those who survived longer would also show signs of encephalitis, that is they became sleepy, lost consciousness or went into a coma, and showed muscular weakness. The unusual feature of this type of encephalitis is that patients develop myoclonic tremors. These are gross, irregular, involuntary movements of the upper or lower extremities. This is typical of brain stem encephalitis, or an encephalitis that involves midline structure at the base of the brain.

Confusion in the minds of the public was further created by a disagreement within the medical community. The Chief of the Section of Epidemic Control of Taiwan's Department of Health, Dr. Wang Li-Shen, a distinguished expert on infectious diseases trained at the University of Kansas, maintained these isolations did not prove that the cause of the outbreak had been found. Enteroviruses are common viruses. The isolated virus might be a fellow traveler, an insignificant contaminant. To show that it is the cause of death in a patient, he maintained, the virus must be isolated from the central nervous system and not just the throat. He further doubted that the sudden deaths could be due to an enterovirus, because he quoted the guidelines of the WHO, which said that hand, foot and mouth disease is a "benign" disease, a disease with essentially no deaths.

On the other hand, Dr. Lin Tsou-yen, chief of pediatrics at the Chang Gun Hospital, also an infectious disease expert, and whose virus laboratory had made one of the virus isolations, maintained that these isolates were significant, and that the microbiological cause of the outbreak had been found.

Enterovirus outbreaks are common throughout the world, including the United States. They usually occur in the spring and early summer in temperate regions. Besides hand, foot and mouth disease, enteroviruses also cause aseptic meningitis, non-specific diarrhea and upper respiratory symptoms. Most of these illnesses are benign and non-lethal. In fact, of the more than seventy types of enteroviruses, only three types of polioviruses cause outbreaks of serious irreversible disease, paralysis due to involvement of the motor centers of the spinal cord, and occasionally death. Polio is well controlled throughout the world by vaccines, including Taiwan. It is on its way to extinction. A non-polio enterovirus epidemic with many deaths was essentially unknown in medicine.

When I went to Taiwan in 1997, I was not planning to work in my main area of research in Pittsburgh, virology. But it was widely known that I was a virologist and it became inevitable that I would be involved in my area of expertise. One of my first jobs was to oversee the effort of NHRI to support a national diagnostic virology laboratory at Cheng Kung University Medical School, in Tainan, in southern Taiwan, whose dean was Dr. Huang Kunyen. Dr. Huang had engaged the help of Edith Hsiung from Yale University Medical School to establish this laboratory. Edith is a world-renowned authority on diagnostic virology. Like me, she is an expatriate born in China. She recommended that one of her students, Dr. Wang Jen-Ren, be put in charge of the laboratory, with Dr. Hsiung coming almost yearly for four or five years to help her. Having her on board assured us of the quality of the laboratory. I have known Edith for years. Her single-minded insistence on quality is the secret of her success in viral diagnostics. An important ingredient is her ability to communicate with clinicians who submit specimens to her. It was no surprise to us that of the dozen virus laboratories associated with teaching hospitals or the Department of Health in Taiwan, it was the most recently established virus laboratory at Cheng Kung University, supported by NHRI, that first isolated Enterovirus type 71 from a throat swab of a patient with hand, foot and mouth disease on May 23, 1998. The second one was made shortly later by the Chang Gun Hospital virus laboratory. These were the isolations I was asked to comment on. After receiving the telephone call in Wellesley, I was back in Taipei in forty-eight hours.

There are seventy-two types of enteroviruses, numbered more or less according to their chronological order of discovery. Enterovirus 71 was discovered only in 1974[85]. It was a relatively "new" virus.

In the beginning, I, like everybody else, was alarmed by the unusual clinical manifestations of those who were hospitalized and who died. Many, including leading virologists, suspected an undetected viral agent, or postulated a cofactor besides Enterovirus 71 to explain the epidemic. One of them was Dr. Ho Mei-shang, who is a physician virologist working at IBMS. One evening at dusk, after work, she and I and Cliff McDonald were walking in the grounds of Academia Sinica, where both NHRI and IBMS were located, discussing this possibility. She told us about her experience visiting patients at Changhua Christian Hospital in Changhua, in central Taiwan, where an unusually large number of patients had been hospitalized and many died. She was horrified at the rapid course and demise of the seriously ill and she was alarmed and puzzled . We discussed a similar outbreak of hand, foot and mouth disease in Malaysia in 1997, where there was also a disturbing number of deaths. They had isolated an adenovirus in addition to Enterovirus 71. Adenoviruses are completely different respiratory viruses, named after the adenoids, from which they were originally isolated. Even if Enterovirus 71 proved to be the main agent, she was of the opinion that there must be a co-factor besides Enterovirus 71, because she did not think that virus alone could explain the whole clinical picture. At the time, many co-factors were suggested in the lay press, such as aspirin, and traditional Chinese medical concoctions of unknown composition, which parents bought and used to relieve the patient's symptoms. Other co-factors thought of were differences in medical care in different regions of Taiwan. Another possibility often mentioned was that this virus was a particularly virulent "mutant" of Enterovirus 71. In fact, the head of the virus laboratory at Chang Gun Hospital, Dr. Shih Shin-ru, came to my office to discuss this possibility. She was convinced that the base (the four chemical building blocks of DNA) sequence of the virus genome would explain its unusual virulence. She wished to ask for funds from NHRI to do the base sequencing of her isolate. I cast doubt on the necessity of this hypothesis. But the base sequencing was later done by a number of laboratories supported by NHRI.

After I got back to Taipei, I drafted a news release for NHRI, expressing the belief that Enterovirus 71 was the cause of the outbreak, and that additional viruses or cofactors were unnecessary hypotheses.

As Dr. Wang Li-Shen maintained, enteroviruses are common viruses that can be isolated from both sick and well patients. But Enterovirus 71 is not one that is commonly isolated in Taiwan. In fact, before 1998, Enterovirus 71 was only isolated in small outbreaks in 1980 and 1986 in Taiwan. Otherwise, it was not found in the records of the laboratories.

In the short time that it had been known, Enterovirus 71 had already produced nationwide epidemics in Bulgaria and Hungary, with deaths, in 1975–1976[86,87]. This was relatively unknown, even among virologists because these countries were then behind the "iron curtain". In Bulgaria, there were 405 cases in a polio-like outbreak, and 44 young children died of "bulbar encephalitis", which we nowadays call "brain stem encephalitis", which is similar to the type of encephalitis we were seeing in Taiwan. The government of Bulgaria, upon the advice of Russian virologists, who were more experienced than local ones, were so convinced that they were dealing with polio that they began a nation-wide immunization program against polio, although the vaccine was already effectively in use. It was only after the Russians isolated Enterovirus 71 from a patient that the program was stopped. These outbreaks proved that Enterovirus 71 is not a "benign" virus, like most other enteroviruses.

Although no cases of pulmonary edema or hemorrhage were described in previous outbreaks of Enterovirus 71, brain stem encephalitis was seen in other outbreaks, most recently in Japan. There was also an isolated case reported from New Haven, Connecticut in 1995, of a child who died of pulmonary edema and from whom Enterovirus 71 was isolated[88].

These are the reflections on which I based my conviction in the news release.

I learned of a relevant but forgotten piece of information later in the outbreak. In the heyday of polio outbreaks, a neuropathologist, A.B. Baker, collected over one hundred autopsy cases of polio patients who had died with "bulbar encephalitis"[89]. There were fifteen cases of pulmonary hemorrhage or edema among these. Since poliovirus can produce pulmonary hemorrhage, it should be no surprise that a related enterovirus, Enterovirus 71 can do the same. He also showed that the poliovirus had affected nerve cells in the "autonomic" nerve centers in the brain stem. This is an area where involuntary nerves that control the blood circulation in the lungs originate. It explains how this group of viruses can produce pulmonary hemorrhage through the nervous system. The lung itself need not be infected. Even old timers like me had not known this. I had missed seeing cases of polio during my training. In these days of information explosion and knowledge advances, one forgets that previous knowledge can be forgotten or overlooked.

I was also willing to accept the idea that nowadays a known virus may cause unusual or unknown clinical manifestations. There are many examples of this nowadays. One that is best known to me is dengue shock syndrome[90]. In the 1950s, William McD. Hammon, my predecessor at the Graduate School of Public Health at the University of Pittsburgh, described shock and hemorrhagic deaths in Filipino and Thai infants due to dengue viruses, a group of viruses that up to that time was only associated with uncomplicated influenza like illness. The dengue shock syndrome has since become a common epidemic problem, not only in south-east Asia, but also in Hawaii and the Carribean.

Subsequent events supported my conviction. An autopsy of a patient with hand, foot and mouth disease who died suddenly with pulmonary hemorrhage was performed at Chang Gun Hospital. He was found to have Enterovirus 71 in the central nervous system. The lead author of the paper was Chang Luan-yin, one of our infectious disease trainees who became one of the leading researchers of this outbreak. Chang, called "Michelle" when she was our trainee, is a graduate of Taiwan University College of Medicine. She is outstanding in perceiving new phenomena and writing them up. The paper reporting this autopsy was published during the epidemic[91].

An analysis of the virus isolations from Cheng Kung University Medical Center and Chang Gun Hospital during the outbreak showed that many enteroviruses were involved[92]. Isolates from uncomplicated hand, foot and mouth disease patients, mostly from the outpatient clinics, were mostly Coxsackie A9 (another enterovirus) and fewer isolates of Enterovirus 71, but isolates from hospitalized patients including those who died were almost exclusively Enterovirus 71. This showed hand, foot and mouth disease was caused by at least two viruses. But Enterovirus 71 alone was found in patients with most of the serious clinical manifestations. Among patients who died and a virus was isolated, 92% revealed Enterovirus 71. It alone was medically important.

Work after the Outbreak

It was important in any outbreak to identify the precise microbiological cause. Finding out the cause of infectious diseases is the major contribution of modern microbiology, beginning with Louis Pasteur and Robert Koch. This is true of individual patients who are sick, and it is true of community outbreaks. The ordinary layman may not understand that this means identifying the exact type of enterovirus. In Taiwan, I had to insist on calling the outbreak an "Enterovirus 71 outbreak" rather than an "enterovirus outbreak" as people in the news media were likely to do. I was upset that in September, 1998, when the dust had settled and all had agreed on the cause of the outbreak, there were still public health professionals who wrote in the newspapers about the outbreak without mentioning the agent, Enterovirus 71. I had to write an article in the newspaper reiterating this point. This is not nit-picking. Without this specific, one cannot begin to talk about prevention and control. Enterovirus infection can be prevented by specific vaccines, which have to be rigorously type specific. This was already shown with the poliovirus vaccines. Three vaccines against the three types of polio viruses had to be developed and manufactured separately before the final product could be pooled into one. Since other enteroviruses are also similar, poliovirus vaccines show that vaccines against Enterovirus 71 is possible and likely to be effective. I favored the development of a killed vaccine of the Salk type[93,94], which is effective and easier to develop than the live Sabin type vaccine[95]. In this respect, my recommendation has not been carried out with dispatch. Six years after the outbreak, a vaccine is still in preparation.

Neither can one talk about epidemic control of all enteroviruses, just like we cannot talk about the control of all respiratory viruses. There are too many and most of them are relatively harmless. One has to talk about control of Enterovirus 71, because that is the one that is medically important. There was already in Taiwan a clinical surveillance system consisting of notification of clinical cases of hand, foot and mouth disease, by a group of primary physicians throughout the island contracted by the Department of Health. This system was originally designed in 1995, to survey cases of upper respiratory infections, diarrhea and dengue. The same group effectively reported hand, foot and mouth disease during the 1998 epidemic. They were the ones who first identified the outbreak. One could attribute that disease to some undefined enterovirus but there was no method to identify Enterovirus 71 clinically. Enterovirus 71 can only be identified by laboratory methods. By monitoring the presence of this virus in the community, one monitors the presence of an agent that may cause death. I proposed that competent virus laboratories take on the job of systematic surveillance of all enteroviruses, which is the first step. The second step is to do the typing for Enterovirus 71, which is more complicated. This proposal was accepted and financed by the Department of Health, and was instituted by the fall of 1998. Taiwan now has a system of laboratory surveillance, not only of influenza viruses, but also Enterovirus 71 and other enteroviruses. This measure has proven important, because there have been almost yearly outbreaks of Enterovirus 71 since 1998.

At NHRI, we initiated a research program to elucidate the epidemiology and serology of Enterovirus 71. I drafted a call for research proposals under the auspices of NHRI. This was done in August of 1998, and by January of 1999, bureaucratic red tape was overcome. Three medical centers, the University of Taiwan, Cheng Kung University and Chang Gun University, were funded to do this research. It was shown that, amazingly enough, seventy percent of the older children and adults in Taiwan were already immune to Enterovirus 71 before the epidemic. Only infants and toddlers six months to three or four years old were susceptible, and they were the ones who seroconverted and became sick during the epidemic. Large families with many young children in rural communities were at highest risk. It became obvious to us that, despite the high level of immunity among older children and adults, a susceptible population of younger children would be available for infection every two years or so. This turned out to be the case, as there were outbreaks of Enterovirus 71 infections in the year 2000 and almost yearly thereafter. Workers at Cheng Kung University also established a mouse animal model for Enterovirus 71 infection. They showed that specific immunity induced by specific antiserum or induced by specific antigen was type specific, as would be expected from what we know about other enteroviruses, such as polio virus. Taiwan became a center of research of Enterovirus 71.

APEC Proposal

In 1997, a year before the Taiwan outbreak, the three parts of Malaysia had a widespread outbreak of cases of hand, foot and mouth disease, comparable in scope to that in Taiwan. They had a similar number of deaths, primarily due to pulmonary edema and hemorrhage. We knew about this outbreak in Taiwan, but we could learn nothing from it. The reason is that the outbreak had no name. No one knew what virus one was dealing with. In 1997, the Malaysian scientists could not agree on the etiology, or cause of the outbreak. This was still disputed among Malaysians three years after the outbreak in 2000, during a conference of APEC nations on Enterovirus 71 sponsored by the Department of Health of Taiwan. The majority in peninsular Malaysia, where the capital Kuala Lumpur is located, incriminated Enterovirus 71, but a minority in Sabah, which is on a large island previously called Borneo, incriminated an adenovirus. We in Taiwan could not benefit from the experience of Malaysia in the spring of 1998. It was only later in 1998, a full year after their outbreak, that their first scientific publication appeared[96].

Many lessons were learned during the Taiwan epidemic. One must establish the etiology of any outbreak as soon as possible. Failing that, one cannot reasonably institute measures for prevention and treatment. Without an agreed cause, confusion is inevitable. I also learned that it may not be possible to determine with 100% certitude the cause at an early stage. Not all the evidence may be in. One might have to venture a calculated guess. Unless this is done in an authoritative and orderly way, disagreements also may contribute to confusion. Despite many experts, Taiwan did not have an authoritative committee which weighed all the data and came to a consensus concerning etiology, prevention and treatment issues. Dr. Wang Li-shen resigned in June, 1998, which cleared the air somewhat. But the Department of Health still failed to assume effective leadership. They could not organize the experts and make them function as a team.

Still, by the fall of 1998, it became apparent that Taiwan had muddled through the epidemic fairly well. It had an excellent system of clinical register of cases by contractual physicians. It isolated fairly early the responsible agent and there was agreement as to the cause. It established a system of laboratory surveillance of specific enteroviruses. And it instituted and completed within six months of the epidemic a series of studies elucidating the epidemiology and pathogenesis of this virus. An authoritative article by a group of Taiwan authors, led by myself, was published in the *New England Journal of Medicine* in September 1999[92].

In view of what we learned and did about Enterovirus 71, Taiwan proposed to the APEC (Asia Pacific Economic Cooperation) meeting of March, 1999 in Hong Kong a "Virus Watch for Children" project. APEC is a consortium of "economies" of the Pacific rim. Most of them are independent nations, like mainland China, Japan, Australia, New Zealand, the Philippines, Indonesia, Thailand, Malaysia, Canada and the United States.

Mainland China allowed two Chinese economies, Hong Kong and Taiwan to participate in this international arena. It is almost the only one that Taiwan can participate in; the other being the Olympic Games. A great deal of attention is given to APEC activities in Taiwan. To have a project proposed by Taiwan accepted by this body would be a boost in Taiwan's international prestige. I was less interested in the international implications of the project than the fact it would be good public health for the entire region to benefit from Taiwan's experiences. So I was supportive of the project. Initially I was given a draft of the proposal authored by one of the staff in the Department of Health. I was asked to edit the proposal. Since I had been a participant of many of the public decisions of the outbreak, and I was in the process of writing a definitive paper of the outbreak with the major leaders who dealt with the outbreak as coauthors, I was in the position of drafting a cogent proposal. I rewrote the entire document, making the following proposals for the APEC nations. Taiwan proposed that the nations conduct a clinical and laboratory surveillance for enteroviruses and Enterovirus 71. Enterovirus 71 activity was not only present in Taiwan and Malaysia, recent outbreaks were also found in Japan, Singapore and Australia. The clinical surveillance and laboratory surveillance that Taiwan instituted was described and offered as a model for other nations. Taiwan offered expert consultation and on-site training on the epidemiological, clinical and laboratory aspects of Enterovirus 71. This was based on the experience that Taiwan workers went through during the outbreak. Taiwan proposed an international data center for Enterovirus 71 located in Taipei, with all APEC nations contributing on a voluntary basis.

The Department of Health accepted my revision and invited me to be part of the delegation to the meeting. Two other individuals representing this project were part the delegation; Dr. Twu Shun-jeh of the Department of Health, and Professor Li Chin-yun of the University of Taiwan Medical School, who is a respected, well known senior virologist. Dr. Twu asked me to present the proposal verbally at the meeting. I did this in twenty minutes, with slides showing clinical and epidemiological data from our outbreak. My talk was the only one that received general applause from the audience, which consisted of delegates from all the Pacific rim countries. Later, it was approved by all the members of APEC, including mainland China. A year later in March, 2000, the Department of Health organized an international conference in Taipei on Enterovirus 71. This was an offshoot of Taiwan's APEC project. The participants from all APEC countries were invited, including Malaysia. It was apparent then that Taiwan had provided a model for monitoring, detecting, diagnosing, management and surveillance of a major epidemic virus. Taiwan made a major contribution to international public health.

chapter

| 15 |

RESEARCH AND ADVOCACY IN ANTIBIOTIC RESISTANCE

In 1996 President Cheng-wen Wu of NHRI expressed the hope that I would come to work there after my retirement from the University of Pittsburgh. I proposed a mission of attacking the antibiotic resistance problem and advocating its solution. I knew that it was a serious problem that had to be addressed by an infectious disease specialist like me. But when I arrived in Taiwan in 1997, I did not know how I could make a difference.

Soon after my arrival, I was visiting with the chief of infectious diseases at the Chinese Medical College at Taichung when I gave a talk there. After telling him of my interest, he said to me,

"Antibiotic resistance is a very, very difficult problem to solve. In the last analysis, it is a governmental problem to be solved by the Department of Health. I really don't know what you can do about it. You should not waste your time on this topic, and you should continue to work in virology, which is your own field."

This candid comment took me by surprise. Chinese are not ordinarily so straightforward, especially to an elderly "authority". At that moment, I directed my resentment at him, because he was an infectious disease expert, but did not consider finding the solution to this problem his responsibility. But I was able to evaluate his comment more objectively later. Coming as it did on top of my lack of a control plan, he discouraged me but he also stimulated me. He alerted me to the importance of the government in any control program, much more so than what I was accustomed to do, coming from the United States. In the United States, one relied much more on the medical societies and voluntary agencies to institute control measures.

Although I had no blueprint for action, I did plan to establish a laboratory to conduct a nationwide surveillance of antibiotic resistance. That was obviously needed research.

I did not know how I could be effective in engaging Taiwan's antibiotic problem in 1997 because outwardly, Taiwan was doing everything recommended by American guidelines. The introduction of compulsory universal national health insurance in Taiwan in 1995 was an opportunity to introduce measures to control antibiotic resistance. The bureau responsible for carrying out this insurance policy, the Bureau of National Health Insurance (BNHI), under the national government and Department of Health, was not unaware of the importance of controlling antibiotic resistance. They adopted a strategy used in the United States and other developed countries, where controlling the use of broad spectrum, more expensive antibiotics (non-first line) in hospitals was their priority. These antibiotics were the second and third generation cephalosporins, moxalactams, fluoroquinolonces, the more expensive macrolids (like azithromycin) and vancomycin. BNHI wrote regulations to restrict the prescription of these non-first line antibiotics in hospitals by requiring further consultation with infectious disease experts or the approval of higher hospital administrators. In order to satisfy the demand for antibiotics, the cheaper, generic, narrower spectrum, first line antibiotics, such as penicillin, ampicillin, methicillin, first generation cephalosporins, erythromycin, and gentamicin were to be "preferentially" used. As in the United States, the focus of attention was on hospital rather than ambulatory practice, because hospital-acquired infections harbored organisms resistant to non-first line antibiotics. The upshot of these regulations was that when I first arrived, the resistance rates in hospitals against these non-first line drugs were holding steady, in fact they had decreased somewhat after 1995. They were on par with the rates in the United States, which were not the best in the world, but tolerable.

I first became acquainted with Taiwan's medical problems in 1992, when I started to go to Taiwan yearly to direct an infectious disease fellowship program. Clinical teaching was part of the program. I listened to case presentations by our fellows in hospitals, and went with them to the wards to see the patients. We also went to the microbiological laboratories to look at the bacteria of patients under the microscope made from smears of sputum and spinal fluid, and we looked at the results of their bacteriological diagnosis and antibiotic sensitivities. A part of patient management required

Staff of Microbial Infections Reference Laboratory (MIRL) or the antibiotic resistance laboratory at NHRI, Taipei, 2000. Fourth from left are Ms. Tsai-Ling Lauderdale, Professor Kunyen Huang and Monto Ho.

data on what antibiotics were effective against bacteria isolated from the site of the patient's disease process, whether it be from his blood, wound, urine, spinal fluid or other body fluids. Going to the laboratory was part of my practice to confirm the laboratory's reports. Coming from the United States, where antibiotic resistance was a recognized major problem, I was not prepared to be surprised. But I was. At that time, over 50% is the Staphylococci isolated in Taiwan were resistant to methicillin, while in the United States, only 10% to 15% were resistant. Gentamicin, a powerful antibiotic effective against Pseudomonas and Enterobacter, was uniformly effective in American hospitals, but in Taiwan, these organisms were 20% to 50% resistant. Taiwan's antibiotic resistance problem was of a magnitude more serious than America's. I was told that practitioners in Taiwan used antibiotics in excess and without discretion. I had noticed that in some cases presented to me, patients with infections received antibiotics not just for a week or two, but for months. They not only received them in hospitals, but also after they were sent home. Part of the problem was cultural. Taiwanese patients liked medicines, especially injectable medicines. Gentamicin was a favorite injectable antibiotic, especially for urinary tract infections. It was heavily abused in ambulatory practice, where it was often given in single injections. Single injections rarely cure any infection. Inadequate doses are a powerful stimulus for development of resistance.

One time I went to the laboratory and asked to see a summary of their antibiotic sensitivity results for the current month. I was struck by a record of vancomycin-resistant Staphylococci. It is well known to clinicians that vancomycin is the last resort against Staphylococci, one that is effective even when methicillin or oxacillin is no longer effective. At that time, no one had isolated a vancomycin resistant strain in the world even though every infectious disease expert was on the look out for one. I asked the laboratory supervisor about her result. She did not blink an eye. She did not think anything of it. She did not suggest it was a possible mistake. Her reaction suggested that she was either ignorant of the facts or she did not care. I believe it was the former. At the minimum, this was indication of inadequate communication and interaction between the laboratory and the clinicians. Later in 2002, when we investigated laboratory practice in Taiwan's hospitals, we found out that most clinicians paid no attention to sensitivity results. They prescribed antibiotics empirically without regard for sensitivity results[97]. This experience told me that there were unique deficiencies in the health care system that contributed to the antibiotic resistance problem.

Another time, I reviewed in a hospital the charts of all patients with infectious diseases discharged during the previous month. I came upon the following case:

A 56 year old male veteran with diabetes was admitted with fever and chills. Two blood cultures were drawn, both of which grew out in a day *Staphylococcus aureus*. This is a common gram positive coccal bacterium with moderate virulence, but which is very dangerous when it causes bacteremia. As soon as the service was aware of the report, he was started on intravenous oxacillin. The sensitivity report of the organism arrived a day later. It was resistant to oxacillin and all other tested antibiotics except vancomycin. The infectious disease service was consulted about the therapy. He was not switched to vancomycin until 48 hours after the report. Despite two weeks of proper treatment, the patient died.

This patient had bacteremia caused by "MRSA", or methicillin (or oxacillin) resistant *S. aureus*. Taiwan's rate of MRSA is so high that the policy of not starting such a patient on vancomycin immediately after admission is already questionable. But because of delay caused by the requirement for consultation before vancomycin could be prescribed, additional delay

Monto with the staff of the antibiotic laboratory at the NHRI in Taipei. Mrs Lauderdale is second from left.

was incurred, so that he did not receive proper treatment for 96 hours after admission. This delay was the probable cause of his death.

Thus, I was familiar with many complex reasons for antibiotic resistance; excess usage of antibiotics, inadequate use of antibiotics, improper usage of antibiotics and inadequate laboratory support for infectious disease practice.

NHRI was very generous in providing funds for purchasing the necessary equipment for a laboratory to do national surveillance of antibiotic resistance, but recruiting the professional personnel was much more difficult. They were either not available or they could not be persuaded to come to NHRI. Most of my staff came from abroad. The first person I recruited was Lan-Lan Yeh, an epidemiologist from the University of Pittsburgh. I was very fortunate in attracting by advertisement Clifford McDonald, a physician who was an epidemic intelligence service (EIS) officer at the CDC (for details see Chapter 17). He was an infectious disease expert and an epidemiologist-microbiologist, ideally qualified for our work. He in turn introduced Mrs. Lauderdale, a Taiwanese-American, who had had twenty years of experience in microbial diagnostics in the United States. With these three individuals, we started the "Taiwan Surveillance of Antimicrobial Resistance (TSAR)". Then, from Hong Kong, I recruited Dr. Siu, who had recently received his Ph.D. from the University of Hong Kong. Dr. Siu is a molecular biologist who had experience in a medical microbiology laboratory at the Queen's Hospital. He had done postdoctoral work on β-lactamase type of antibiotic resistance in Hong Kong. The only local scientist we were able to engage was Dr. Lo, who was a molecular biologist who did her postdoctoral work at MIT on the virulence of Candida, the most common fungus that can cause disease.

The work on antibiotic resistance done in Taiwan up to this time was limited to studies from separate hospitals or medical schools. TSAR was designed to survey the entire country in an epidemiologically meaningful way, so that a comprehensive picture of antibiotic resistance would be obtained.

In 1998 Dr. Yeh and I visited more than fifteen different hospitals throughout the island, in order to enlist their participation in TSAR. We met the top personnel of each hospital who were working in administration, in the microbiology laboratory, in infection control, and in infectious diseases. We were always politely received, coming as we were from a new prestigious institute. The fact that I was a member of Academia Sinica commanded a great deal of respect. But when the chips were down, it was not easy to convince them to contribute to a cause in which their immediate gain was obscure. I could only detail the national importance of the resistance problem, and ask for their cooperation.

We eventually were successful in recruiting forty four out of a total of sixty-five major hospitals of Taiwan. This effort resulted in our collecting, by the end of 1998, 6,254 laboratory isolates of bacteria from these hospitals for our laboratory. Each isolate was worked up, in terms of its bacteriology, and its antibiotic sensitivities against ten to fifteen different antibiotics. This stupendous job was performed under Mrs. Lauderdale's direction with the help of five technicians in the course of about six months. The first round of TSAR was completed in 1998 (TSAR-1)[98], and the second, third and fourth rounds (TSAR-2, TSAR-3, and TSAR-4) were completed in 2000, 2002 and 2004.

The Biology of Antibiotic Resistance

How did antibiotic resistance become such a major problem in the world? Alexander Fleming, a British microbiologist, had discovered penicillin as early as 1928. By chance, a house mold, a penicillium, had contaminated a blood agar plate on which he had grown colonies of disease producing *Staphylococcus aureus*, a round shaped bacterium that stains positive by Gram stain, and caused bacteremia in our patient. He noted that the colonies around the mold were lysed; that is they had dissolved, or were faded and dying. This signal observation led to his discovery that a living organism could produce a substance that could kill bacteria, and he coined the term "antibiotic". But he could not produce the antibiotic in a purified, clinically useful form suitable as human medicine. This had to wait until World War II, when Howard Florey and Ernest Chain purified penicillin and convinced the U.S., government and a group of pharmaceutical companies to produce it on an industrial scale. Initially, all available penicillin was restricted to the armed forces. But by 1944, enough was made so that it became available to the public.

In retrospect, penicillin was the first and the best antibiotic discovered. It was effective against most of the important types of pathogenic Gram positive bacteria, such as Staphylococci, Streptococci, Pneumococci, and diphtheria bacillus. Some important Gram negative bacteria such as the major causes of meningitis, Meningococci and Hemophilus were also cured. The major causes of venereal diseases, Gonococci and syphilis, could both be cured in a patient in one sitting by a single shot of penicillin! Even more remarkable was the fact that penicillin was not toxic. It produced no side effects except occasional allergy. Another asset of penicillin was that it was "bactericidal" rather than "bacteriostatic". This means that it actually killed bacteria, rather just stopping their multiplication. The killing action is necessary for the complete cure of some serious and difficult infections, such as bacterial endocarditis, or inflammation of the heart valves. When, in the late forties, Selman Waksman discovered streptomycin, which was broadly effective against other Gram negative bacteria not touched by penicillin and also against the tubercle bacillus, we thought that the millennium had arrived. In the fifties, with the availability of additional groups of antibiotics, the consensus was that the infectious diseases had been conquered. One just had to make antibiotics available to all who needed them.

Sixty years later, penicillin is no longer miraculous. It is no longer entirely effective against any of the organisms cited above, with the exception of Streptococci and syphilis. Streptomycin is no longer in general use. This has happened because of developing antibiotic resistance. Fleming had predicted that it would be a problem.

"The greatest possibility of evil...is the use of too small doses so that instead of clearing up infection, the microbes are educated to resist penicillin and a host of penicillin fast organisms is bred out which can be passed to other individuals and from them to others until they reach some one who gets a septicemia or a pneumonia which penicillin cannot save"[99].

Either excessive or inadequate use of antibiotics can promote resistance. A few years after the introduction of penicillin, it stopped being effective against one of its major targets, the Staphylococci. Due to excessive use of penicillin in hospitals and clinics, and the ubiquity of penicillin in patients' bodies, a strain of resistant Staphylococci, that had the resistance gene against penicillin, a gene that codes for an enzyme that destroyed penicillin, penicillinase, was selected to grow in preference to the usual penicillin sensitive Staphylococci. They gradually predominated, first in hospitals and later in the entire community.

I recall the days I was in training at the Boston City Hospital in the 1950s. Our wards were constantly threatened by patients who had infections due to penicillin resistant Staphylococci. They had to be quickly identified, and isolated since these organisms were readily transmitted to other patients and the staff. An entirely new medical subspecialty was developed in the United States, to prevent and care for "hospital acquired" or "nosocomial" infections. For patients with life threatening septecemia and pneumonias due to the penicillin resistant Staphylococci predicted by Fleming, our infectious disease service, under the leadership of Max Finland, provided combinations of less effective antibiotics for treatment. They were bacteriostatic rather than bactericidal drugs and were less effective than penicillin.

Those crisis days were finally overcome by the discovery of a class of novel, semi-synthetic penicillins that were resistant to penicillinase, methicillin and oxacillin. These antibiotics could cure staphylococcal infections due to resistant organisms, and they had all the assets of penicillin. For a while it seemed that human science and technology would always stay ahead of the development of resistance. But today in Taiwan, about 80% of the Staphylocci cultured from hospital patients are resistant to methicillin and the related cephalosporins. In the United States, the figure is about 50%. The only remaining bactericidal antibiotic we have against Staphylococci is vancomycin. Two years ago, vancomycin resistance was reported in Japan, and it has now been reported in the United States.

The story of Staphylococci is paralleled by that of Pneumococci, the most important cause of pneumonia. The exquisite effectiveness of penicillin against Pnemococci was maintained for forty years. The first resistant strains showed up in the eighties in Spain, a country until recently noted for its excessive use of antibiotics. From Spain, the non-susceptible Penumococci gradually conquered the whole world. When I first went to Taiwan to train infectious disease fellows in 1992, penicillin non-susceptible Pneumococci were a curiosity. Now, the Pneumococcus that can be cultured from the throats of normal children in daycare centers in Taiwan are 80% non-susceptible. The sad part of the story is that resistance has also developed against other unrelated antibiotics, which ordinarily would be alternatives to penicillin.

The versatility of bacterial resistance has become a legendary story in bacterial genetics, a body of knowledge that has progressed along with progression of bacterial resistance. All types of bacterial resistance are due to inheritable genes. We can no longer look at resistance genes as unusual mutants on ordinary chromosomes. Mutants occur about once in a million divisions. Even among bacteria which can multiply every twenty minutes, mutants are relatively uncommon and take time to appear. What we have is the phenomenon of "transferable resistance", heredity not mediated by traditional chromosomes. They are in mediated by "plasmids", a small collection of genes outside of chromosomes, or by "transposons", small pieces of DNA that can jump from one piece of DNA, such as a chromosome, or a plasmid, to another piece. Plasmids and transposons are exchanged between bacteria when they have sex, that is when they conjugate. The remarkable thing is that they do not have to belong to the same species. Resistance genes can be exchanged between individual bacteria of different species. The picture we have now is that of a bacterial world, like the one in our large intestines, which contains billions and billions of individual bacteria belonging to hundreds of bacterial species, forming a vast interactive microbial world, exchanging plasmids and transposons in response to changing environments, such as one containing antibiotics that have being ingested by the human, where the resistant strains are selected and grown out. The diverse genetic mechanisms made the development of resistant bacteria a more common occurrence.

The story of Staphylococci and Pneumococci can be repeated with other pathogen-antibiotic pairs. There is a separate story with each pathogen and an antibiotic to which it has become resistant. After we have reviewed dozens of these stories, we realize that practically all the bacteria that cause our infectious diseases and all the antibiotics at our disposal adhere to one relentless law. Any bacterium originally susceptible to an antibiotic will eventually become resistant to it. The only question is when.

If use of antibiotics is inevitably accompanied by the development of resistance, is there anything we can do about it? The basic facts are clear. Resistance only develops in an environment in the patient where both bacteria and antibiotic are present. The antibiotic selects out the resistant strains in this environment. If we stop using antibiotics, resistance would not develop. The more antibiotics we use the higher the rate of resistance. Not only that, since generally resistant bacteria are generally less fit to survive than non-resistant bacteria, resistance will regress or be reversed if the antibiotic is not used. But obviously we cannot stop using antibiotics, because they are life saving drugs. They are essential for the practice of medicine. Under this constraint, what we can strive for is to delay development of resistance and try to keep it low. We promote proper and economic use of antibiotics and we try to prevent the spread of resistant bacteria. Proper use of antibiotics means that they should only be used when they are effective, and they should be used in doses that will cure the infection, not more or less. The major difficulty in enforcing this rule is that many infectious diseases cannot be diagnosed as soon as the patient is seen. Our clinical and laboratory know-how is not that good. Very often, antibiotics have to be used empirically, without hard evidence. Even admitting this necessity, there are still vast areas of medical practice where antibiotics are grossly misused. Abuses can be corrected given proper education and willingness. A primary example is the widespread use of antibiotics for upper respiratory infections, where the pathogens are usually viruses, which are insusceptible to antibiotics. But there are many others. There is a lot of room for promoting proper and economical use of antibiotics.

The simplest and most often effective way to limit the spread of antibiotic resistance, is recognizing the carriers of resistant bacteria in hospitals and institutions and isolating them so they cannot transmit their bacteria to others.

The best evidence that these measures of constraint are effective is the great disparity between different countries in the western world. Culturally similar countries have different rates of resistance. For example, the resistance rates are fairly high in the United States, but they are significantly lower in Canada. They are high in Belgium, but they are low in neighboring Holland and the Scandinavian countries[100].

Precise reasons for these differences are complex, but they can all be boiled down to differences in medical and institutional practice. One area not in the medical world relevant to antibiotic resistance is worth noting. This is the large amounts of antibiotics used in food animals.

Veterinarians noted after antibiotics were discovered that minute amounts of antibiotics added to animal feed dramatically promoted weight gain in animals. This discovery had tremendous economic implications, since it reduced the turnover time of livestock farming as much as 50%. The scientific explanation of this phenomenon is unclear. Pretty soon all farmers throughout the world, including farmers in underdeveloped countries, started adding antibiotics to the feed of their chickens, pigs and in fish ponds. Right now worldwide, at least half of all antibiotics are for veterinary use. Low levels of antibiotics in the intestines of animals where these bacteria grow have became a hotbed for the development of resistance. Many bacteria of food animals can cause diseases in humans; these are called "zoonoses". Examples are Salmonella and Campylobacter, major causes of outbreaks of gastroenteritis. Resistant Salmonella and Campylobacter developed in animals have been shown to be transmitted to humans. Right now in the United States there is serious concern that Campylobacter resistant to fluoroquinolones, an important new class of antibiotics against which resistance has not yet developed has been transmitted to human from food animals. The FDA has posted a discussion of the proposal that these drugs should be banned in veterinary medicine.

The northern European countries have pioneered the development to monitor and restrict the use antibiotics as feed additives and disease prophylaxis in animals. In 2001, Denmark became the first country to eliminate antibiotics as additives for growth. The government provided a fund for any farmer who suffered economic loss because of this measure. At the end of the year, no claims were made. Additives may no longer be effective in this day of modern, sanitary farming, perhaps because the high level of antibiotic resistance already developed in animals may have offset their effectiveness. The reason is unclear. But measures adapted in Denmark cannot be readily adopted in the United States. The United States is a larger and more complex country, where there is a very strong farm and drug lobby.

The Effect of TSAR-1

On July 24, 1999, we organized a meeting of workers from all our participating hospitals as well as guests from the medical community, infectious disease community, and from governmental agencies, to participate in a conference where we announced the research results from TSAR.

This was a festive day. Our main guests were middle-level personnel; technicians from hospital microbiology labs, nurses and personnel of hospital committees on hospital-acquired infections, and the infectious disease physicians of Taiwan's sixty-five medical center and regional hospitals. We intended to reward those who had helped us by submitting specimens, but we also invited those who did not. The session was educational. Besides results of TSAR, other research of our group was also reported. The day was interrupted by a luncheon. Every participant received a "bien dan", which is a delicious, hot boxed lunch consisting of Chinese-style meat or fish, vegetables and rice.

The result of TSAR-1 in 1998 along with other studies showed that Taiwan had at least three major high level resistance problems; the resistance of *Staphylococcus aureus* to oxacillin and first generation cephalosporins, the resistance of Pneumoccus to penicillin, and the resistance of aerobic gram negative rods to aminoglycosides (gentamicin). In terms of degree of resistance of these antibiotics, they were among the highest in the world. Oxacillin (similar to methicillin), penicillin and aminoglycosides were all first line antibiotics. In terms of mass or amount of antibiotics used, penicillin and oxacillin, first generation cephalosporins and aminoglycosides are the most frequently used antibiotics. They are among the so-called "first line antibiotics", or antibiotics that have been designated by the BNHI as common, cheaper narrow-spectrum antibiotics. Our TSAR results highlighted the importance of the resistance against first line antibiotics. The high rate of resistance being developed against them by important pathogens threatened their continued usefulness[98].

The importance of our finding was that it showed that Taiwan's approach to the antibiotic resistance problem was wrong, or at least misplaced. Taiwan had copied the approach of the United States and western countries. It concentrated on the resistance of the more expensive, broad spectrum non-first line antibiotics, and neglected resistances against the first line antibiotics. In fact, in the regulations of BNHI, first line antibiotics were to be "preferentially" used, a statement copied from similar guidelines in the United States. What happened was they were preferentially abused.

This conference received a great deal of media publicity, and it attracted the attention of the officials of the Control Yuan. The Control Yuan is a unique ombudsman branch of the Nationalist or Taiwanese government, historically Chinese in origin. Its sole purpose is to search for and document transgressions of governmental agencies, and enforce corrective measures. A group of three Control Yuan officials, under the leadership of Mr. Chang Teh-ming, beginning with materials that we provided from

TSAR, went on to study the whole antibiotic problem in Taiwan and to document it. Mr. Chang and I became good friends. He is a lawyer, with no medical or scientific training. But, partly as a result of frequent conferences with me, he became an expert. After an investigation of about six months, they came out in February 2000 with a document, really a published book, called *Abuse of Antibiotics is Harmful to the Health of People in Taiwan*[101]. The Control Yuan concurrently issued an official directive, which had legal power. It directed the responsible governmental agencies; that is, the Department of Health (DOH), the Council of Agriculture (COA), and the Department of Economics (DOE) to correct abuses in the areas of antibiotic regulations.

They compiled a compendium of abuses by these agencies. For example, they pointed out that the three agencies did not have precise data concerning the amount and type of antibiotics that were imported or manufactured. As a matter of fact, the DOH, COA and the DOE used different terms for the same antibiotics, and uniform data could not be collated. The COA did not enforce its rules concerning the type and amount of antibiotics permissible as animal feed additives. The COA did not have methods sensitive enough to detect all antibiotic residuals in meats. There was no clear distinction between antibiotic additives given for nutritional or for medical reasons. The Control Yuan officials pointed out regulations that were in the books but were not enforced. For example, they used data provided by us to show that rules of the BNHI, which is an agency of the DOH, restricting use of antibiotics for prevention of wound infection in clean surgeries to three days, were almost uniformly violated. This was substantiated by data provided by us[103].

After the DOH and COA received orders to correct abuses, they sought my advice, knowing the role we played in the Control Yuan document. Many sections within DOH were concerned with one or another aspect of the corrective measures, such as the sections dealing with drug policy, infectious disease control, or food sanitation. In order to coordinate the efforts of different agencies within DOH, the director of DOH put Dr. Chang Hong-ren, the assistant director of DOH, in charge. Dr. Chang consulted me frequently and we became good friends. I observed that he as well as other officials within DOH evolved from being reluctant officials carrying out an unpleasant assignment to eager advocates promoting the rigorous execution of measures to control antibiotic resistance. This was to me a gratifying change.

A National Project against Antibiotic Resistance

I felt official action by the Control Yuan in April, 2000 could correct some abuses or inadequacies of individual regulations, but that strategy alone could not address the heart of the problem of antibiotic resistance. They were piecemeal solutions, very much like the eclectic efforts of the Infectious Disease Society of Taiwan to educate physicians about the proper use of antibiotics.

The heart of the problem was that Taiwan needed a national program to reduce the yearly progression of antibiotic resistance, or at least stop its progression. This could be done by a drastic reduction of consumption of first line antibiotics. I felt that was the only way that the resistances could be reversed or at least retarded. This was the concrete, overall action against antibiotic resistance I had been looking for since 1997. This realization was epiphany for me.

Reduction in antibiotic consumption to address the resistance problem is not the usual strategy in the United States. I had not considered it in the beginning. One usually concentrates on guidelines for proper antibiotic usage, and isolation procedures for patients with resistant organisms. But limited experience in northern Europe has shown that reduction of consumption can work[102]. In 1992, a high rate of resistance of Streptococcus against erythromycin, around 50%, was found in ambulatory patient clinics in Finland. This came from the preference of physicians for an oral medication in preference to penicillin, which had to be injected. A nationwide program was launched to promote the use of penicillin instead of erythromycin. Erythromycin usage dropped. The resistance rate against erythromycin began to fall within a year and was reduced around 10% in two years. This showed that consumption can be reduced if we specify what, where and how it should be reduced, and reduction can lead to reversal of resistance.

On June 30, 2000, at my instigation but under the chairmanship of Dr. Huang Kun-yen, NHRI called a national conference to discuss a national program against antibiotic resistance. For this meeting, members of the Control Yuan, medical academia, and relevant governmental agencies were invited.

My keynote address at this conference pointed out that in order to reduce antibiotic resistance, it was necessary to reduce consumption of antibiotics. To do this effectively, one must find areas where there is a great deal of consumption of antibiotics and where it is also abused. I proposed two specific projects to reduce antibiotic consumption for Taiwan. One was to rigorously enforce the BNHI regulations concerning use of antibiotics for prophylaxis against wound infections in clean surgeries in hospitals. The other was to restrict the use of antibiotics for upper respiratory infections in outpatients.

About 25% of antibiotics used in hospitals are for surgical prophylaxis. Dr. McDonald and I published a paper describing the practice in Taiwan's hospitals[103]. This was based on a study of a thousand patients in 1999 who underwent "clean surgeries", that is surgeries in which all tissues contacted were "clean", or free of bacterial contamination. These are the most frequent procedures in hospitals. They include herniorrhaphies, thyrodectomies, joint replacements, vertebral disc removals, coronary bypasses, hysterectomies, etc. Studies in the United States and Canada had earlier determined that the source of wound infections after surgeries, is bacteria from the patient infecting an open wound during the surgical procedure. Prophylaxis was effective if there was an effective level of antibiotics in the patient's bloodstream while the surgical procedure was in progress. This is for the duration of only a few hours. One or two intravenous doses are sufficient, given half an hour before beginning of surgery. Taiwan's regulations of BNHI written in 1995 are in accordance with these guidelines. Our research results showed that 70% of the patients received prophylactic antibiotics for more than three days, the outer limit set by BNHI. About half off them received the antibiotics later than half an hour before the procedure. Often they were given them after the operation. This paper was evidence of a fact suspected, but not proven, that written rules were flagrantly violated.

Upper respiratory infections are usually caused by viruses and not by bacteria; use of antibiotics is rarely justified. To document Taiwan's practices in this area, with the help of our statistician Agnes Hsiung at NHRI, we turned to the database of BNHI. Here is evidence of the marvel of modern technology. Ninety-eight percent of Taiwan's population is covered by universal health insurance, and the medical record of each insured individual is in the data base. After working for a week, we were able to generate the needed data. There were, in 1997, about 60 million ambulatory patient visits for upper respiratory infections among the population of 23 million. Thirty percent of the patients received antibiotics. There were annually 2,850 ambulatory patient visits per thousand people in the population for upper respiratory infections. By comparison, the comparable number in the United States was 133. Thus there were about twenty times more patient visits in Taiwan than in the United States for upper respiratory infections. Since thirty percent of the patients in Taiwan received antibiotics for URI, the tremendous amount of antibiotic usage in Taiwan is apparent. Almost all of this usage was "first line".

Results of the National Project

The two projects I proposed for reduction of antibiotic usage were eventually accepted by the Department of Health (DOH) and the BNHI. During the latter half of the year 2000, DOH, under the leadership of Dr. Chang Shan-shwan, who is director of infectious diseases at the University of Taiwan Medical School and a friend and supporter of our program, undertook an exhaustive survey of the use of prophylactic antibiotics for surgeries in each of Taiwan's hospitals. Materials gathered were used for education of surgeons in proper usage. The BNHI came under the new directorship by Dr. Chang Hong-Ren by the end of 2000. He was the previous assistant director of the Health Department, with whom I had worked closely to carry out the directives of the Control Yuan. He was fully aware of my opinion that what the Control Yuan ordered was not enough to solve the problem of antibiotic resistance. He was aware of my recommendation to control the use of antibiotics for upper respiratory patients. Just two months after his assumption of his new duties, BNHI announced that beginning February 1, 2001, the regulations concerning reimbursement for antibiotics used for upper respiratory infection would be changed; that is, no reimbursement would be allowed unless there was evidence of a bacterial infection.

After the new regulation, total ambulatory antibiotic consumption was reduced 40% in each of the following years, 2001, 2002 and 2003, compared with 1999[104,105]. In addition, resistance to erythromycin (one of the antibiotics whose consumption was reduced) by Group A *Streptococcus pyogenes* diminished in three Taiwan medical centers[106] and in TSAR isolates from the entire island[107] in 2002, 2003 and 2004. These are results similar to what was found in Finland when consumption of erythromycin was reduced. They attest to the effectiveness of reducing resistance by reducing antibiotic consumption. We were anticipating such results.

Antibiotic Usage in Animal Husbandry

Before I came to Taiwan, I knew that excess usage of antibiotics in the agricultural area was a global problem, but I had little expert knowledge in this area. Our newly recruited colleague, Cliff McDonald from the CDC, had done research work in this area and was very knowledgeable. Before the action of the Control Yuan, we had completed a small survey of antibiotics used in animal feeds, which showed wide use of antibiotics as feed additives[108]. In my consultations with the Control Yuan, I had emphasized to them the importance of looking at antibiotic usage in agriculture as part of the problem of antibiotic abuse in Taiwan. Therefore, their report on this issue mentioned prominently the importance of controlling antibiotic usage in animal feeds.

Shortly after the COA received the directive to institute corrective measures, Mr. Sung of COA and Professor Lai of Taita (University of Taiwan), both veterinarians, visited me to discuss possible countermeasures to satisfy the directives of Control Yuan. At that point, I had been studying up on the problem of antibiotic abuse in agriculture and came across the name of Dr. Frank Aerestrup of Denmark. From his papers, I knew that he had instituted an antibiotic resistance surveillance program among food animals in Denmark and had published some results of his studies[109]. It occurred to me that he had done more in a small country such as Denmark than people in countries like the United States, which has difficulty in instituting surveillance and control measures because of its size and complexity. When these gentlemen came to see me, I presented to them a paper by Dr. Aerestrup on his surveillance work. It became apparent to me only during the discussion that these veterinary authorities in Taiwan were totally unaware of the method by which they could try to understand the problem of antibiotic abuse and how to control it. The idea of undertaking a surveillance program was completely novel to them. Because of this, I invited by e-mail Dr. Aerestrup to come to Taiwan, in order to investigate the situation in Taiwan and to educate the veterinary and agricultural community.

With Mr. Sung's help, I arranged a 10-day visit for Dr. Aerestrup during which he toured chicken and pig farms and slaughterhouses in Taipei, Chang-hua and Tainan counties, and he gave talks about the use of antibiotics in food animals and the way they had approached the problem in Denmark. He noted during the tour that the Taiwanese regulations of use of antibiotics in animal feed were imprecise, and these were not adhered to. He noted the inadequate role of governmental agencies in the microbiological diagnosis and care of infectious diseases in animals. He gave three public lectures that explained what he had done in Denmark and indirectly indicated what was to be done in Taiwan. His conclusion was that Taiwan had a severe problem of abuse of antibiotics in animal husbandry, and that there was need for a surveillance program to try to understand the extent of this problem. In addition, he suggested that Taiwan establish, like Denmark, a zoonosis center where the medical and veterinarian authorities could interact and arrive at a common position in fighting the problems of zoonotic infectious diseases and abuse of antibiotics in animals. Dr. Aerestrup's visit had a profound impact on the authorities of COA, and members of the academic veterinary community.

Immediately after his visit, the COA contracted Professor Lai of Taita and us at NHRI to undertake a national surveillance of antibiotic resistance in food animals. This program was carried out with unusual dispatch, within months. The bacteriological work was done by Mrs. Lauderdale in our laboratory. We isolated *Escherichia coli*, enterococci, Salmonella and Campylobacter from fecal specimens of chickens and pig farms in three different counties of Taiwan, collected by personnel of the governmental agricultural clinics. Antibiotic sensitivities were conducted with the isolates. This work was completed in six months. Unfortunately, a companion survey of antibiotic usage was unsatisfactory, partly due to the reluctance of the farmers to provide information. At this point, we know that there is a tremendous amount of resistance to most of the first line antibiotics in the animal flora in Taiwan, just like in human beings, but it is unclear where and how the vast amount of antibiotics is being used. The COA has not been forthcoming in collecting data on antibiotic usage. Still, the COA banned the use of avoparcin, an antibiotic related to vancomycin used in humans, and some other antibiotics, in 2000. This is in consonance with what was done in Europe in recent years. However, other problems are pending, such as the problem of fluoroquinolones. We found significant resistance of gram negative bacteria against fluoroquinolones, and extremely high resistance against nalidixic acid, a non-fluorinated quinolone. The latter signifies prevalence of quantitative resistance against fluoroquinolones. We noted that the U.S. FDA is presently considering a proposal to ban the use of flouroquinolones in animal husbandry. Unfortunately, the limits of bureaucracy had been reached. Nothing has yet been done about this problem.

It is apparent that after the action of the Control Yuan, the DOH and COA both responded quickly with action. This was very satisfying. Nevertheless, I feel it was not enough.

Further Work to be Done

Antibiotic resistance is a long-term problem that will be ever present as long as antibiotics are used. There are countries in the world where antibiotic resistance is extremely low and very well controlled, such as the northern European countries and Canada. Most of the rest of the world, whether developed or undeveloped, suffer from various degrees of antibiotic resistance. The problem in Taiwan is accentuated by certain cultural characteristics. We have already mentioned the excessive number of visits of ambulatory patients in Taiwan. Physicians in Taiwan suffer from an excessive demand for attention from patients. To accommodate them, physicians frequently spend only three to five minutes or so for a single ambulatory patient. Clearly, one cannot do a thorough diagnostic workup under such time constraints. One might ask, why do patients in Taiwan put up with five-minute visits with their physician? The answer is that their primary purpose in seeing a doctor is to get a prescription. It is easy to give the patient a prescription in five minutes. This excessive demand for medicines is cultural. In the Chinese conception, every illness requires some sort of medicine. The idea that some diseases do not require medicine is unacceptable. This is the main reason for the tremendous excess of antibiotic prescriptions. The system is made worse by reimbursements of physicians for all patient visits, and of pharmacies for all prescriptions by health insurance. In the future, the whole system for ambulatory patient care needs to be reformed.

Physicians have more time for proper diagnostic workup of patients in hospitals, but we have found that antibiotics are still not used according to microbiological diagnosis, but by empiric judgment[97]. Excess usage is the norm. What Taiwan needs are rigorous, empiric guidelines concerning every indication of antibiotic use. We started an intensive study of lower respiratory infections, particularly bacterial pneumonias, which are the most important and numerous infectious diseases in hospitals in Taiwan. There are few studies of important causes of pneumonia in Taiwan. I feel there should be more studies of hospital practices in Taiwan. Taiwan's medical community should underwrite their own practice guidelines, rather than abiding blindly by what is done in the United States and other foreign countries.

Taiwan should be constantly reviewing its ambulatory and hospital practice areas to look for areas of gross antibiotic abuse. Reducing antibiotic consumption in these areas, as we suggested in two areas in 2000, can lead to reversal of current levels of resistance. This is the most effective strategy to solve the antibiotic resistance problem. Taiwan should adopt this strategy as a longstanding, national policy.

In the agriculture area, it is important to recognize the powerful effect of the pharmaceutical drug lobby, which exerts undue pressures on the farmers. There must be enough research to make clear what antibiotics are being used and where. Such information, together with data of antibiotic resistance of animal flora, will make clear the effects of antibiotic use and abuse. Then there must be an open discussion among governmental officials and experts in academia concerning what should be done about problems created by antibiotic usage. What I am concerned about is that this orderly sequence of events is not being followed, and the whole problem is still closed to informed public scrutiny.

chapter
| 16 |

IMPROVING TAIWAN'S MEDICAL TRAINING AND MEDICAL EDUCATION

Years of research and teaching in an American university have given me some insight and understanding of America's success. For the same length of time that I have been in the United States, the United States has been the most important and influential foreign country for Taiwan. Taiwan has been protected militarily by the Untied States. It is literally the only Western country to which Taiwan looks for guidance, not just politically but culturally. It has sent thousands of students to study in the United States and seeks to simulate the United States. I found that coming from the United States, and having studied and worked there, was an advantage in my work of trying to help Taiwan. I did this in areas where I had some experience; in Taiwan's medical training programs and in Taiwan's education. Still, what I found is that Taiwan belongs to a different culture, and simulation must be selective and thoughtful in order to succeed. I encountered unexpected problems which required tailor-made solutions.

Infectious Disease Training Program

The Chinese, including Taiwanese, are conscientious students. They work hard and they complete their assignments. So what can go wrong in a training program in infectious diseases for physicians?

I completed training for thirty-five physicians in four classes, each class consisting of two years of training, from 1992 to 2000 before I found out.

In Chapter 13 we mentioned that infectious diseases was one of the three categorical areas we wished to develop in Taiwan, along with cardiovascular diseases and cancer, when we planned for establishment of the Institute of Biomedical Research (IBMS) in the eighties. One of the major problems we encountered was lack of personnel in research leadership to staff IBMS. Since there were no leaders, we went about training them. Paul Tso pioneered in the formation of a training program in oncology. This was the beginning of the specialty training in oncology in Taiwan, where it is now an established subspecialty of internal medicine.

It was assumed that I would lead in the development of an infectious disease training program, especially since I had run an infectious disease training program for physicians for many years at the University of Pittsburgh. But I went about this project carefully, because infectious diseases was already a medical specialty in Taiwan, and there were already a number of training programs in existence in teaching hospitals. In fact, there was an "Infectious Diseases Society of the Republic of China" just like the "Infectious Diseases Society of America". It acted like a subspecialty board and it examined and certified physicians in this subspecialty. What Taiwan lacked was not clinical specialists in infectious diseases, but research leadership in infectious diseases.

The introduction of another infectious disease training program met with some resistance from some of the specialists, despite the strong outside support of the Secretary of Health, Chang Po-ya. It could not have been carried out without the support of a group of leaders among the specialists, led by Hsieh Wei-chuan, chief of infectious diseases at University of Taiwan Medical School.

In 1992, Dr. Wu obtained the support of the Department of Health, IBMS and the Infectious Disease Society of Taiwan to establish, under my direction, an infectious disease training program for physicians. Our training program was similar to the American program in American medical schools and teaching hospitals. After completing a residency in internal medicine or pediatrics, the candidate fellow receives two years of training in infectious disease practice, lectures and a research dissertation. For the clinical and research training, we enlisted the participation of four major teaching hospitals in Taipei; Taita (University of Taiwan) Medical School and Hospital, the National Defense Medical School and Tri-Services Hospital, the Veterans General Hospital, and Chang Gun Medical School and Hospital. Each hospital had a well developed infectious disease service with categorical patient beds available for teaching. For research training, the teaching staff of the infectious disease services were enlisted and reinforced by the teaching and research staff of the departments of internal medicine and pediatrics, clinical microbiological laboratories and the pathology departments of each hospital.

For direct teaching, I enlisted a faculty of six or seven visiting professors in infectious diseases from the United States for each class. They came to Taiwan for a period of one to two weeks of full-time instruction. They were given lodgings in one of the four hospitals where the teaching took place, and where they could be close to the patient beds.

Right from the beginning, I was acutely aware of the deficiencies in Taiwan's educational system and tried to correct them. While students get good grades and collect the desirable number of diplomas, the entire educational system, including medical schools, is geared to passing examinations and getting good grades. School children are trained to regurgitate what they hear from teachers or read in their books. Little attention is paid to developing the individual student's ability to think and solve problems. In order to enter desired high schools and colleges, most students also go to "cram schools" after regular school in order to sharpen their ability to pass examinations. These vicious habits, cultivated in a cultural environment where academic achievements are so highly prized, are one of the reasons for the lack of innovative researchers in Taiwan.

Our directed teaching activities had two characteristics to offset these habits. Visiting professors were full time teachers while they were in Taiwan, so they had maximum possible contact and interaction with the fellows. The number of fellows in each class was kept at a minimum; five to fifteen in each of our five classes during the eight years of existence of the program. Second, we emphasized in our teaching problem solving, discussion, and individual reports by fellows rather than direct lecturing by the professors. Cases and papers from the current literature were brought in by the professors and fellows to be discussed together. Fellows were given individual assignments related to the cases or papers discussed, to work up and report. They had to do background reading in standard texts or in classical papers. This type of teaching and learning requires a great deal of effort of participating fellows and instructors. In order to establish the proper tradition, I was always the starting visiting professor to start each new class. A fellow had the following comments about me in our newsletter of the training program.

"Dr. Ho is like an old eagle. He oversees every aspect of our program. I realized that I had to work hard to prepare for classes and in order to get the most out of the discussions. So much so that I lost two pounds and I started to realize the joy of our studies in the second week. I was feeling improvement constantly."

Forty-two physicians were trained in five classes from 1993 to 2001. Of the numerous visiting professors I had invited from the United States, those who participated for more than three terms besides myself included Dr. Chien Liu and Dr. Chung-cho Cho of the University of Kansas Medical School, who were respectively the chief of the Division of Infectious Diseases in adult medicine and in pediatric medicine. We had in addition Dr. Calvin Kunin, who was retired chair of the Department of Medicine of Ohio State University Medical School and Dr. Richard H. Michaels, who was from the University of Pittsburgh, where he was previously the chief of the Division of Infectious Diseases in the Department of Pediatrics. Dr. Tom Chin was another faithful visiting professor, who gave to each of our five classes a mini course on medical epidemiology and biostatistics. He was previously chair of the Department of Preventive Medicine at the University of Kansas. Besides these veterans, others came for one or two sessions, and some of them made deep impressions on our fellows. They were Dr. Don Armstrong of the Sloan Kettering Medical Center of the Cornell University School of Medicine in New York City, Dr. Lowell Young of San Francisco and Dr. Victor Yu of the University of Pittsburgh. Dr. Yu came once for the second class in 1994. But he established personal relationships with many of the fellows and continuously collaborated with them in research. This collaborative effort with some of them has lasted more than ten years, from 1994 until the present time (2004).

Cal Kunin was another visiting professor who established close relationships with our fellows and continued to have interest in their development. Additional relationships added to his involvement with Taiwan. He was a member of the advisory committee of our research group at NHRI and he took an active interest in our work in antibiotic resistance after I came to Taiwan in 1997. My relationship with him goes way back, as we were both infectious disease fellows in the laboratory of Max Finland of the Thorndike Memorial Laboratory of the Boston City Hospital in 1956. He became renowned for his work on the renal clearance of antibiotics and urinary tract infections. He has been interested in one or another aspect of antimicrobial research during his entire career. As President of the Infectious Disease Society of America from 1986 to 1987, he alerted the American and the global medical community to the impending danger of increasing antibiotic resistance.

One of my initial concerns concerning the training program was that fellows would have difficulty finding and completing a research project in the two years they were with us. Indeed, the first class had some difficulty in carrying out this assignment. I brought up this problem to the curriculum committee of the training program chaired by Professor Hsieh Wei-chuan of Taita, who was then the veteran chief of infectious diseases there and one of the most respected persons in the field. Without his active support of the program, it would not have been feasible. He discussed this problem with his colleagues in the medical centers that formed part of our consortium. Beginning with the second class, each fellow was able to find a research project with one of the teaching faculty, execute it, and complete a research report. Most of these reports were subsequently published in medical journals in Taiwan. From 1997 to 1999, twenty-nine papers were published with twenty-six of our fellows as authors or co-authors. At that time, we had trained a total of thirty-seven fellows, thus 70% of our fellows then became authors. This would seem to have been a very satisfactory result as far as research training was concerned.

Indeed, the general opinion seemed to be that we had been very successful in training infectious disease specialists. But in the spring of 1999, I decided to undertake an anonymous evaluation of our training program among our trainees.

All of the responses expressed gratefulness for the training they had received, but some of the critiques were illuminating as well as surprising. Even though we had always emphasized the importance of active learning, as opposed to passive lecture-type learning, one critique mentioned that there should have been even more active problem solving-type learning. Although we had convinced him of the importance of this type of active learning, he still thought we did not have enough. I bring this up because in retrospect, I think he was right. This type of teaching and learning is hard work! Even instructors from America may not do enough of it. Hopefully this lesson will benefit them further in their own continuing education.

Another surprising comment was that they wanted more participation of the visiting foreign professors in their research training. I had assumed that they were satisfied with the research training, as I was, at least superficially. I took this critical comment to heart though and invited Cal Kunin to join me in a thorough review of the ongoing research projects of every one of the seven fellows of the 5th class. Cal came to Taiwan twice for this exercise.

Each interview took three to six hours. We meticulously went over the nature of the project, its objectives, and the methodology of research of each fellow. What we found was that each fellow took a ready made project from his mentor and did what the mentor directed. So the progress of the work was rapid, and results were forthcoming, and the writeup was relatively straight forward. What we found missing was the meat of the research.

There was no agonizing over the precise phrasing of the problem, the deliberation of the research methods to be used. They avoided agonizing over the objectives and purposes of their projects or the difficulties and problems concerning the methodology. They did not do enough thinking about their project to write an in-depth discussion of the significance of the findings. They did not appreciate that agony was essential for effective learning, especially in research. Their projects were, in a way, tailor-made by their mentors. In our interviews, we discussed with each fellow his/her research project in the light of these headings, and we considered additions, improvements and modifications that could be made in their projects.

It finally became apparent to me why the fellows were able to produce a research report in such a short time, as opposed to my experience in the United States. This is a general comment on the cleverness of the Chinese. They are very good at perceiving and taking shortcuts, avoiding the slow, plodding approach of more careful and compulsive people. It is slipshod. It is not adequate training for a good scientist.

After training 44 fellows in five different classes, we decided to terminate the training program. We had not intended originally to make this training program a permanent one. There were already ongoing infectious disease training programs in various teaching hospitals in Taiwan. What we were trying to do was to demonstrate what we thought were more effective training methods, and to train a cadre of good researchers. I believe we have succeeded in doing that. Two major obstacles also blocked the continuation of the program. First, we could not offer an advanced degree, like a Ph.D. degree. Ph.D. programs have in the last ten years become extremely popular in Taiwan, even in clinical departments. But they can only be offered by a university, and not NHRI. Second, we could not offer our trainees any financial support. The cost of training had been borne by the hospitals that provided the fellows. Theoretically, they were still in the employ of these hospitals during their fellowships, and one of the perennial problems had been that they were called to perform service while they were in training with us. From a personal point of view, this program gave me an intensive opportunity to teach, somewhat like my sabbatical in Australia, a comparable intensive opportunity to do personal research.

The Peer Review System of Research Grants

A typical American innovation is the peer review system of research grants. It satisfies the democratic need for recruitment of experts from various levels of the research establishment, avoiding the "buddy" system, where an oligarchy of established investigators with special interests control all major decisions. It is a system that promotes the best science in the long run. Investigators are spurred to write a grant that best expresses their capabilities, by the promise of the best impartial scientific decision and open competition. The decision is accompanied by a critical but constructive written critique available to the applicant. This system, not dissimilar to the review of manuscripts by the best scientific journals, has been adopted and is continuously improved by the extramural research program of the NIH, which is arguably the largest and best medical research establishment in the world. This system is the clue to the superiority of American medical research in the last half century.

Even before the establishment of NHRI in 1996, we already instituted an extramural research program in Taiwan which adopted the American type of review system. We recruited a panel of expert reviewers from the United States who had firsthand experience with the system. We frequently encountered the problem that applicants for a research grant did not know how to write an application that reflected their capability. I remember giving a public lecture in Taiwan in 1992 on the ingredients of an adequate research application. That was also the year when NHRI first called for research applications from Taiwan's medical schools and research institutions. The number of applications increased from twenty-three that year to 127 in the year 2000. The amount of research dollars allotted rose from 1,300,000 to 300,000,000 NT (1 US Dollar = 35 Taiwan dollars). Applications were reviewed in one of three committees — medical sciences, bioengineering, or public health. For three years, I chaired the medical science review section. From 1998 to 2000, I was chair of the council of the entire review system. The reviewers were recruited from various fields in the medical sciences, mostly in the United States, but more recently also from Taiwan. Each application had a primary and secondary reviewer. Their critiques were written up and distributed by a parent committee before a general meeting.

On a designated date, the three review committees met and reviewed each application and the primary and secondary critiques. After discussion, a consensus decision was reached on the score of each application. A consensus review was drafted supporting the score and returned to the applicant. Although only about 20% of the applications could be funded, this consensus document contained the constructive criticisms and recommendations of the committee. Very often, suggestions were made for improvement of the grant so that they could return with a revised or improved application during the next round. The document was meant to be educational rather than officious. This comprehensive and rather expensive system of review has become a model in Taiwan, and it itself has become a stimulus for scientific progress. It is in sharp contrast to previous reviews, which were frequently short, perfunctory, uninformative and redundant.

Evaluating Taiwan's Medical Education

During my sojourn in Taiwan, I was interested in all aspects of Taiwan's educational system. I had the opportunity to participate formally in evaluating institutions of medical or paramedical education. The first opportunity was in March 1999, when I was appointed to a committee under the chairmanship of Dr. Liang Kung-Yee to evaluate the Institute of Epidemiology of the School of Public Health of the University of Taiwan. Then, in the year 2000, I became a member of the "Taiwan's Medical Accreditation Council" (TMAC), upon the recommendation of its chairman, Dr. Huang Kun-yen. In 2001, I participated in the evaluation of the University of Taiwan Medical College. In April 2002, we evaluated the Taipei Medical College and the Fujen University Medical College. As an example of my critiques, I would like to report on my impressions of the two institutions within the University of Taiwan.

The University of Taiwan is a unique institution of higher learning in Taiwan. It is the oldest and has the greatest prestige. Whatever shortcomings I may have to report on should not be considered unique to the University of Taiwan, but probably represent a general phenomenon, with the situation being very likely worse at other institutions.

What I found surprising at the Institute of Epidemiology of the University of Taiwan was the lack of mentoring of the junior faculty. This is probably more or less the case throughout Taiwan. This is different from what I had anticipated, and seems uncharacteristic of a Chinese cultural setting. One would have expected the department head or the professor in Taiwan to have a great deal of prestige within the department and to be the chief mentor of all the junior faculty. This is certainly true in the United States, where the department head is not only the administrative leader of the department, but is also the academic leader. He has the responsibility of overseeing the intellectual output and the creative development of the junior faculty. In contrast, the department head in Taiwan is much less authoritative. He is frequently an elected head whose term is limited usually to 3–5 years. In the United States, a department head is responsible for the recruitment, development, and promotion of his faculty. Younger faculty members in Taiwan are frequently left to their own devices. The department head in Taiwan has much less flexibility in all personnel matters. Practically every administrative procedure, including faculty promotions, is predetermined by a set of rigid rules, and exceptions are unusual.

During our evaluative visit, we interviewed every faculty member at the Institute. I participated in interviewing a young biostatistician who became a specialist in a rather narrow area of biostatistics. He had obtained his Ph.D. in Taiwan. In response to our question of whether he had any dissatisfactions, his response was that he was not promoted rapidly enough. We looked at his bibliography and it was apparent that he was a hardworking researcher with quite a few publications. We asked him whether he had considered going abroad to meet and study with colleagues in the same field. His surprising response was that he had no need to meet people with the same interests. The fact that an aspiring young faculty member would deny the importance of communicating with scholars with the same interests must indicate a blind spot in his development. One of the basic requirements for progress in research is communication and discussion with people of similar interests. A basic mentoring process within the department should have made this apparent to him.

Actually, I have found that there is lack of effective communication at various levels of society throughout Taiwan. Students, scholars, and scientists do not gather together to discuss their work often enough. This may be blamed on a basic tenet of Confucian culture. Silence is golden and is valued as a sign of deference. There is therefore indirect discouragement of frank thoroughgoing discussions. This may be the case even in Chinese families. Even there, frank and thoroughgoing discussions may not be encouraged. When I was director of the clinical research division in NHRI, I had to go out of my way to encourage discussion among the scientists and staff, for both science and personnel problems.

TMAC's evaluation of the University of Taiwan Medical School took five days. Four days were spent on on-site visits at the campus. Such thorough evaluations were a first for Taiwan. They follow a model of evaluation practiced in the United States and Australia.

The University of Taiwan Medical College (Taita) aspires to be the best in Taiwan in terms of quality and abundance of facilities. On the other hand, like the three other national medical colleges, it is entirely supported by the government whose funding has to be equitably distributed. In this respect, Taita is like medical schools on the European continent. I find it strange that Taita does not resort to private sources in order to maintain its superiority, as a medical college would do in the United States. The idea of funding by large voluntary and private donations has not taken hold. Taita has a long and illustrious history. It has a vast group of medical alumni, many of whom are extremely well to do, but they have not been educated to give to their alma mater. I suspect that Taita would be very successful if this source were better tapped.

Looking at Taiwan's medical education from an American's point of view, I found that the most important difference was in clinical training. The difference is in "observation" versus "participation". Bedside teaching in Taiwan's medical schools is largely limited to passive observation of patients, while in the United States, the student gradually participates in patient care. Assuming responsibility is a powerful teacher. "The most important thing about patient care is to care for the patient," said Francis W. Peabody, the late, great Harvard professor of medicine[79]. This attitude still pervades in American medicine, while it is less apparent in Taiwan and China. These aspects of medical education do not come with material and technological advances, of which Taiwan has an abundance. They are part of a culture that is less easily taught.

Taita has had recently the laudable tradition of allowing a number of students to study in foreign medical schools during their sixth, or clinical, year. During the evaluation, we had access to the reports of these students. There were twelve such students in 2000. Most of them went to the United States, but a few went to medical schools in Japan, or in Germany. From their reports from such places as the Brigham and Women's Hospital in Boston or the University of Rochester, I could see that language was not a major problem, except in the beginning. These students were able to participate in the work of the fourth year medical students in the United States. These students were first struck by the high degree of enthusiasm for patient care that the American doctors showed. This enthusiasm was evident from the visiting physician to the resident house staff and the student clerks. The second point noted by these students was that they, as student clerks, were part of the patient care system and not, as in Taiwan, "foreign bodies" who were largely ignored by the more senior physicians. Very often, the student clerk was the first person to see a new patient. These student observers confirmed my impressions of the present status of clinical education in Taiwan.

The essence of effective clinical training is assumption of responsibility. Effective training can never be accomplished by mere onlookers. How to provide student clerks and the junior resident staff the chance to assume responsibility is a difficult problem in Taiwan. Part of the problem is cultural. The Taiwanese system is basically authoritarian, where the responsible physician is either the primary resident physician or the professor. The absence of thorough, democratic discussion of case management is also part of the culture. Because of these difficulties, improvement is difficult, despite the fact that leaders in medical schools are aware of these problems.

There is no doubt that in Taiwan, Taita is at the forefront of medical research. The number of papers published by the faculty in international journals with high "SCI scores" is comparable to the record in the United States. Part of this is due to the upsurge of knowledge and expertise in molecular biology at Taita and other medical schools since 1980. Still, I notice a deficiency in clinical research. There are case reports, case series, and summaries of data from diagnostic laboratories, but there is lack of clinical research of importance and significance. There are not enough clinical trials. In order to have this type of research, it is necessary to have an adequate infrastructure. There must be excellent hospital records where full clinical details are available and recorded. Physicians have to be eagerly interested in their cases and concerned about their patients' therapy. Taita should consider the example of American institutions like the Mayo Clinic, where clinical research is part and parcel of its excellence in diagnosis and therapy. One must not merely concern oneself with publications in basic science journals with high SCI scores. Some excellent journals of clinical investigation, like the New England Journal of Medicine, and the Journal of Clinical Investigation, have received relatively few articles from Taiwan.

Partly due to the influence of Dr. Huang, there has been an upsurge of interest in humanistic studies in Taiwan's medical curriculum. What has happened, however, as in so many other things, is to take the easy way out. The easy way to provide humanities is to give the medical students lectures on various humanistic aspects, such as medical ethics and various aspects of medical sociology. This is inadequate. In order for humanistic education to succeed, one must approach it like good clinical education. There has been a tendency to spoon feed what has been taught in the West. I believe problem solving is a more important approach than passive lecturing. Many problems in the social sciences have no absolute answers. What the student needs to do is to appreciate the existence of such problems, and they should be taught to seek out their own answers. Students should be exposed to indigenous problems in Taiwan, such as problems of its universal health insurance and its outpatient system, and problems of relationship between the physician and patient. Actual problems of medical ethics seen in practice locally should be discussed rather than theoretical guidelines. One of the important objectives of humanistic studies is to impart to the student a sense of responsibility to solve some of these problems. Admittedly, this approach takes more effort and resources. But there is no easy way. After all, students easily construe humanistic studies to be outside of their field of interest. It has to be carefully cultivated in order to succeed.

part

IV

THOUGHTS AND REFLECTIONS

chapter

| 17 |

REFLECTIONS IN TAIWAN

Diversity and Scientific Excellence

One of the major problems I encountered after I arrived in Taiwan in 1997 was the difficulty in finding competent investigators for the research program I envisaged. I needed someone who was a physician investigator, knowledgeable in diagnostic microbiology, understood infectious disease and versed in the problems of antibiotic resistance from the medical, epidemiological, as well as scientific perspectives. In February of 1998, we advertised in the medical and scientific journals for such a person. One day while at home in Pittsburgh, I received from my office in Taiwan an e-mail about someone who was responding. I was able to call him conveniently immediately in Atlanta. Cliff McDonald introduced himself as a medical graduate of Northwestern University, who was board certified in internal medicine and in infectious diseases, and who also had a fellowship in diagnostic microbiology. He was at that moment undergoing one year of training as epidemic intelligence officer at the CDC. His special assignment concerned the areas of antibiotic resistance, hospital-acquired infection, and diagnostic microbiology. His work had taken him to various foreign countries, such as Brazil, Africa and mainland China. He also told me that he had spent a summer in Taitung in Taiwan as the resident physician working with outpatients in a Christian hospital. He was very interested in going back to Taiwan to work. After hearing all of this over the phone, I could not believe my ears. Here was a person made to order for what we had in mind. I immediately invited him to come to Pittsburgh to meet me.

The day of our conversation was a Thursday, and I had intended to return to Taiwan on the following Sunday.

I said, "When can you come to Pittsburgh to see me?"

Cliff said, "Let's see, I'm pretty tied up next week......"

"I don't mean next week, I mean this week. How about the day after tomorrow, Saturday?"

"......It turns out I am free on Saturday. I'll come".

I met him at the Pittsburgh airport. Cliff is a tall, lean young man with an easy smile, and a ready approachability. I brought him to our home in Pittsburgh, and we spent the morning together. Carol prepared a Chinese lunch for us. Cliff decided after our conversation to join us in Taiwan, and he would bring his whole family with him.

The apparent ease with which Cliff made his decision misled me as to how difficult it is to recruit foreigners to work in Taiwan. Was Cliff an exception? In order to move to Taiwan, he had to sell his house in Atlanta, and he and his wife brought a family of four little children, ages 2 to 8, with them. All his children were adopted. This is a reflection of their Christian dedication. Six months later, I met them at the airport in Taipei upon their arrival in December, 1998. In order to accommodate their small mountain of luggage, I had to bring, besides a private car, a van that could hold nine people. Everyone in the department helped them to settle down.

My secretary Elsa Ting in Taiwan was particularly helpful. She helped them find a spacious modern apartment, located a kindergarten and primary school for the children, located for them a teacher of Chinese, and helped them with problems, such as healthcare and insurance, etc. Cliff himself went to work right away, like a duck to water. He was more familiar with antibiotic resistance and diagnostic microbiology than I was. Cliff was not only knowledgeable. He had innovative ideas and could write research papers. During his sixteen months in Taiwan, he became involved in six research papers. He is an outstanding product of American medical education and the effectiveness of the epidemic intelligence service (EIS) training program of the CDC. This was the brain child of Alexander Langmuir, famous medical epidemiologist, who created the program originally to allow highly trained physicians to serve the country by becoming "shoe leather" (ground level) epidemiologists at the CDC instead of being drafted in the army. The continued popularity and effectiveness of the program survived the military draft, and has become a permanent institution of the CDC.

The recruitment of Cliff McDonald demonstrated to me that diversity was important for scientific progress. A visit to Singapore convinced me.

On November 16, 1997, while I was at NHRI, I received an invitation from the director of the Singapore Institute of Molecular Biology, Kris Tan, to attend the celebration of the tenth anniversary of his institute. I have already mentioned that Kris was a research fellow with me in 1970–1972 (Chapter 8)[46]. During the two-day celebration, we learned about its accomplishments. In ten years, more than 500 papers were published by the institute and fifty-one doctoral students were trained. There was an equal emphasis on pure science and applied technology. In 1995 and 1996, the important British journal *Nature* recognized and lauded the institute.

Because of their similarity, I could not help comparing Singapore's institute with Taiwan's Institute of Molecular Biology. As discussed in Chapter 13, Taiwan's institute was founded about the same time as Singapore's institute. The two institutes are similar in size and almost identical in mission and objectives. Taiwan's institute, like the Institute of Biomedical Sciences (IBMS) of Academia Sinica, has achieved a great deal and is state-of-the-art in many areas of research. On the other hand, in direct comparison with Singapore's institute, I felt that the latter was a notch ahead.

Both institutes were led by ethnic Chinese. As is well known, Singapore is a tiny city-state with a population of merely two million. Although 70% of its population is Chinese, Malays and Indians are the other official minority constituents, making up the rest of the population. The Chinese in Singapore and Taiwan, like Chinese everywhere, are heavily influenced by the Confucian heritage. That is, there is a strong emphasis on education and scholarship. The main difference between Singapore and Taiwan is that Singapore has also adopted a tradition of diversity and internationalism, while Taiwan, by comparison, has remained more indigenous.

This was already apparent from the guests invited for the celebration. At such celebrations, both in Singapore and in Taiwan, there are usually a number of guests from the United States or from Europe. On the other hand, what struck me, coming from Taiwan, was that in Singapore, there were many guests from neighboring countries, like Indonesia, Malaysia, and the Philippines. One would hardly see any guests from these countries on a similar occasion in Taiwan. This made me feel that Taiwan, situated as it is on an island in Southeast Asia, is more insular and regionally isolated.

The recruitment of research personnel in Singapore is not limited by race, belief, sex, or citizenship. The only criterion is scientific excellence. Among the thirty research investigators, eighteen were ethnic Chinese, seven came from the mainland, five from Singapore, three from the United States, two from Hong Kong, and one from Taiwan. Ten were Caucasians, of whom four were from the United Kingdom, two from Germany, and one each from the United States, Canada, Australia, and New Zealand. Finally, two were from Southeast Asia, that is, one each from India and Malaysia. In contrast, the research personnel of Taiwan's institute was almost exclusively ethnically Chinese. And most of the Chinese were born in Taiwan.

We already described the history of recruitment of research personnel for IBMS (Chapter 13). After some initial difficulty, the Institute of Molecular Biology, like IBMS, succeeded in recruiting senior Chinese American scientists from the United States to return to Taiwan. This success was due to the establishment of adequate research facilities, and the development of a competitive pay scale in the eighties due to Taiwan's economic prosperity. This policy has perhaps become too successful. In order for Taiwan to compete successfully on the international scene, it is now necessary to introduce more diversity, along the lines of Singapore.

Why is it not possible to maintain scientific superiority by recruiting only indigenous scientists? History provides us with the answer. In the end of the 19th century and beginning of the 20th century, Germany was a leader in all the sciences, including the medical sciences. Although she suffered defeat during World War I, it did not affect her scientific leadership. On the other hand, after Hitler's accession to power in the 1930s, a doctrine of racial purity was introduced. Hitler thought that the excellence of German science was solely due to the racially pure Germans. The exclusion, exile, and eventual extermination of Germany's Jews, many of whom were accomplished scientists, set back German science and universities such that it has not been able to recover its leadership role, even long after World War II.

Another example of the inadequacy of indigenous science is Japan. Although defeated in World War II, Japan has become economically a superpower. Scientifically, it has also achieved world stature. On the other hand, it has not been possible for Japan to supersede the United States. The main reason is that the United States is more diverse than Japan. Japan has difficulty exceeding the limitations of its own culture and language. The situation is quite different from the Germany of Adolf Hitler. Japanese culture is extremely conservative. They have not been able to utilize even their own human manpower adequately. Women are not adequately utilized. It is extremely difficult for non-Japanese to become a part of Japanese society. The Japanese themselves realize this problem. Japanese research institutions have recently attempted to diversify by getting out of the constraints of conservative institutional practices, and introducing English as the working language. But progress is slow and difficult.

The obstacles to diversification in Taiwan are similar to those in Japan. There is, in Taiwan, among some people, a misguided movement to prioritize its indigenous characteristics. On the other hand, Taiwan has had fifty years of active importation of American influence. This should be an example that could favor the promotion of diversity.

Diversity in the promotion of scientific excellence has an important requirement: The promotion of English as the working language. Since World War II, English has gradually become the universal language throughout the world. Nowhere is this more apparent than in the sciences. Linguistic chauvinists such as the French have had to yield to English as the working language in international meetings and even in their own journals. The Japanese, too, are publishing their most important articles in English international journals. A main problem in Taiwan is that despite years of instruction in the English language in the secondary schools and colleges, English has not yet become a working language. That is, an ordinary research person is not able to understand, write, or speak English with ease. This, of course, is not the case in Singapore, where throughout society, including its Institute of Molecular Biology, English is truly the working language.

My later experience in Taiwan showed that I was initially overly optimistic about introducing diversity in Taiwan's scientific establishments. Yes, Cliff McDonald was an exception. He left after only eighteen months, although he had a five year contract. The reasons for his premature departure are complex. Prolonged residence in a country where an unfamiliar foreign language is customary, concerns about health insurance, the frequency of typhoons and earthquakes, and the general inadequacy of amenities for foreigners are all reasons. It is quite possible that Cliff would have stayed longer were they in Singapore. Still his eighteen months were a significant contribution to NHRI. He has maintained his contacts with Taiwan.

Academia Sinica, under the leadership of Lee Yuan-tseh, who in the beginning advocated indigenous science, is now at the forefront of promoting diversity in scientific research in Taiwan. It is introducing English as the working language, recruiting students and staff from abroad, and improving the environment for foreigners on its campus. He has become a firm believer of diversity and excellence. But progress takes more time than I had anticipated.

Officialese

Taiwan is a country of paradoxes. On the one hand, people strive to be efficient, Americanized, and modern. On the other hand, many things get bogged down, projects are left unfinished or incomplete, and there is obvious inefficiency. Soon after my arrival, I was struck by the prevalence of "officialese", and the profusion of official documents circulating among Taiwan's bureaucracies, including the one I was working in. Officialese and official documents are ancient characteristics of Chinese governmental culture. After all, China is the longest surviving state, and China invented many of the encumbrances of "red tape". Officialese, according to Fowler, is "the language characteristic of officials, or of official documents". I was so struck by this situation that I wrote an article that was published in a local newspaper.

When I first arrived in Taiwan to work in an official organization, what struck me was that official documents seemed to be the lifeline of institutions in Taiwan. In the beginning, I could not understand what "official documents" meant. I had not been familiar with this term after working as an administrative leader in an American university for more than twenty years. We had very little use for "official documents" in doing business in the United States. Official documents such as contracts and other legal documents were rarely confronted. On the other hand, in Taiwan, I discovered that my desk would be piled high with "official documents" every working day of the week. Clearly, we were not dealing with official documents as understood in the United States.

What is used in the United States to communicate with one's colleagues? When I first arrived in Taiwan, I asked Elsa, my secretary, for stationery on which to write a "memo". In the United States, memos take the place of many official documents in Taiwan and are used to communicate with colleagues at all levels. Elsa told me that memos are not written in Taiwan. There was no stationery on which to write it. There were only "official documents". I was totally flabbergasted. I gradually realized that "official documents" are primarily documents exchanged between the upper and lower levels within a bureaucracy. By and large, these are directives from the superior level, or "requests" from the lower level. Such documents are highly formalized and written in a special language called "officialese". Nothing can be accomplished without them. This leaves, however, a striking deficiency in communication among parallel levels.

An example will illustrate the importance of official documents. Prior to coming to Taiwan, I submitted to NHRI what I thought was a precise and detailed plan for the establishment of the "Microbial Infections Reference Laboratory" to carry out my project on antibiotic resistance. This plan was approved by the president of NHRI, and I thought that this was a fulfilled condition of employment. Upon my arrival, however, I discovered that this was not the case. In order for such a plan to be carried out, every step must first be translated in official documents written in officialese. One prerequisite of such documents is that they must adhere to the many rules and regulations governing budgetary and personnel practices. For example, I requested in my original plan six to eight technicians for the laboratory. I only realized after arrival that this request was against the personnel regulations, since a "P.I". (principal investigator), which I was, was entitled at NHRI to only one or two technicians and six or eight were out of order. It took renegotiation with my superiors to legitimatize this exception. Naturally, at the time I considered these additional procedures nothing but "red tape", impeding progress.

Memos in the United States and official documents in Taiwan are not always equivalent. Memos are more informal, and the hierarchical structure of the bureaucracy is not ostentatiously emphasized. Agreements reached through memos in the United States have the effectiveness of being official. In Taiwan, such agreements must be legitimized by formal "official documents". In fact, theoretically, anything done at all must be supported by an official document. That is the reason why there is a profusion of such documents on one's desk. In the old days when material goods were less abundant in Taiwan, requests for pens and pencils or stationery had to be approved through official documents. Each had to go through three or four different levels in the bureaucratic hierarchy. Once the request was approved by the highest authority, then more official documents are needed to disburse needed funds. This involves several steps. First one has to obtain permission to budget the request. After that is obtained, one then requests the actual funds. One can imagine the size of the bureaucracy to support all this "red tape".

It can be seen from the above account that many of these official documents are uninteresting and a waste of time. They are all written in officialese, which is a style of writing approximating the ancient Chinese literary style, noted for its conciseness and obtuseness. Very often an official document would land on my desk, and I would have trouble figuring out what the original problem was. The issue frequently involved complicated regulations which are never quoted or explained in the document. Needless to say, lots of time was wasted in going over these documents. One cannot simply dump the lot in the waste basket, because a vital issue may be hidden in one of them.

Couching the official documents in officialese is one of the primary problems of these documents. I believe it is actually an impediment to effective communication. Learning how to write in officialese is an art and skill demanded of office workers throughout Taiwan. I read an announcement in the newsletter of Academia Sinica that a training class was being held for the teaching of officialese for new employees. Ironically, I had to recall that a former president of Academia Sinica was Hu Shih, who in 1919 started the great literary renaissance movement in China, the primary objective of which was to promote the use of the vernacular as opposed to literary Chinese. How would he regard the promotion of literary Chinese in Chinese officialese, which is not only prospering in Taiwan, but is also used in mainland China?

Stylized officialese is actually inimical to clear cut solutions of problems. It often restricts communication, especially between parallel agencies, rather than facilitating it. I submit that promoting vernacular Chinese would improve understanding of these documents and reduce their mass. I provide an example to show that official documents not only impedes efficiency, but it may even be a life and death matter.

Early in the administration of the new president Chen Shui-bian in 2000, his prime minister Tang Fei complained about the volume of official documents he had to go through every day. On July 22, 2000, there was a flash flood in the county of Chiaye. Four men were stranded on an island in the middle of a rising river. They watched the rising waters for three hours about to inundate them, waiting for help. Despite the awareness of the local provincial and central government of their plight, and the availability of rescue helicopters, none was forthcoming in the three hours of waiting, and the four men were drowned. One can imagine the profusion of official documents required for the dispatch of helicopters. Their demise was attributed to red tape.

Working at NHRI and attending the weekly meeting of the bureau chiefs, I became convinced that procedures in Taiwan are encumbered by too many detailed regulations. This is part of the classical bureaucracy. For example, the number of regulations governing the purchases of equipment is overwhelming. The original objective of these regulations was to stem corruption. In fact, what it does is to increase red tape and reduce efficiency. We spent a great deal of time at our conferences going over the language of rules and regulations. They had to be constantly revised as new contingencies come up. The main reason is because they were too detailed to begin with. I would like to propose that major resolutions made at the chiefs' conference be immediately carried out without drafting additional documents, but this is contrary to custom. Every resolution has to be translated step by step into official documents going through prescribed channels of the hierarchy. This is the basic reason why NHRI, like many other bureaucracies in Taiwan, is burdened by an excess of administrative personnel.

I realize my views on officialese and official documents flies in the face of ancient tradition. It is said that the term "red tape" was introduced by the British in India, where documents were bound by a red tape. This practice, however, was said to have originated in China during the Chin dynasty. On the other hand, the practices of the Chin dynasty were less than two hundred years ago, but they fall back on a history almost 2000 years of governmental and bureaucratic practices.

Whither Taiwan?

When I first arrived in Taiwan in 1997, I thought that I, as an outsider, should keep my views on the future of Taiwan to myself. This was reinforced by my perception that Taiwan's politics was not freely discussed among my local friends and acquaintances. This, however, did not prevent me from trying to learn about this subject by reading newspapers of different political persuasions and books to enhance my understanding of Taiwan's past and present.

I was actually more occupied by learning more about Taiwan as part of Chinese culture. It will be recalled that one of my motives for going to Taiwan was to live among Chinese and work within Chinese culture. My thoughts and reflections in Taiwan focused more on its Chinese nature, rather than its uniquely Taiwan aspects.

The election in March 2000 of Chen Shui-bian as president, who was a frank proponent of independence for Taiwan, forced upon my attention the controversy between proponents of "One China" and "One China and One Taiwan". Discussion of this controversy became much more common after his election. As a result, I wrote an article for the pro-independence Liberty Times called "One China". The newspaper accepted it for publication, but published it side by side with another article called "One China and One Taiwan", written by a Professor Chen of Chun Shan University. I accepted this as a good democratic compromise.

Mainland China has stipulated that only after overtly accepting the principle of "One China" can talks between mainland China and Taiwan be resumed. They argue, and many in Taiwan also believe, that this common premise was assumed by both sides before the election of President Chen. This basic assumption, however, was destroyed in 1999 by the previous President Lee Teng-hui shortly before his retirement. He announced the point of view during an interview with a West German broadcasting company that China and Taiwan were "two nations and two states". He suggested that the two states should deal like East and West Germany, on an equal basis and negotiate its unification from that posture. This point of view created a great stir, and an instant rebuke from mainland China. Most people believe that the two states theory is simply another way of advocating Taiwan's independence.

The biggest problem with ex-President Lee's view is that it is not supported by any other country, not even Taiwan's main ally, the United States. The United States has agreed repeatedly with declarations with mainland China that Taiwan is a part of China. This is the irrefutable result of World War II. It was confirmed in 1972 when President Nixon and mainland China signed the Shanghai Declaration. The division of Korea and Germany into communist and non-communist parts was also a result of World War II, but it was recognized by international usage. President Chen also recognizes the dilemma of the pro-independence movement in Taiwan. Since his election, he has not actively advocated Taiwan's independence. "Independence" seems to be a uniformly avoided word in the free press in Taiwan, and in public debate. Nevertheless, President Chen has not overtly accepted the concept of "One China". Therefore relations with the mainland remain in a deadlock.

Return to China, or Return to Taiwan?

When I decided to go to Taiwan to work, I was thinking that it was equivalent to returning to China. Would I have gone to Taiwan if I had known that Taiwan would become independent? This is a question brought up by Mr. Chen's election in my mind. I had gone directly from mainland China to the United States more than fifty years ago. My studies and my career have been in that country. Unlike many other mainland Chinese, I had not lived or studied in Taiwan. Therefore, my relationship with Taiwan was scant and superficial. Still, when President Wu Cheng-wen of the National Health Research Institutes invited me to go to Taiwan to work in 1997, I had no hesitation whatsoever. Going to Taiwan to work was in my mind the equivalent to returning to China. To be eventually of service to China had always been one of my life's aspirations. This has been so for more than fifty years.

A Nation and a State

The controversy about Taiwan's independence is unfortunately made more knotty by a deficiency in the Chinese language. In Chinese, the word for a "state" and the word for a "nation" is the same. It is also confounded by the fact that in the last century, there have been three major changes of the political authority in the Chinese "state". The first was the change from the Ching Dynasty to the Republic of China in 1911. The second was the proclamation of the People's Republic of China on the mainland in 1949. The third was the acquisition of political authority in Taiwan from Japan after World War II by the Republic of China, which survives there as its former vestige. The controversy concerns the sovereignty of the latter. The definition of a nation can best be understood by looking at history. That there was a German nation was already apparent in 800 A.D., when Charlemagne was proclaimed emperor of the Holy Roman Empire. However, it took a thousand years for the German nation to become recognized as the German state. This happened in 1871 when the Prussian king William I became the emperor of the German empire after defeating France. Hence, it is quite apparent that nations can exist apart from statehood.

The biggest difference between China and Germany is that the Chinese nation has been independent as a state since the first emperor Chin Shih Huang Ti more than two thousand years ago (221 B.C.). Chinese culture is not the most ancient in the world, but the Chinese state, with its ups and downs, is the most ancient one. The long survival of the Chinese state is also a unique characteristic of Chinese culture.

I would like to give a personal experience of nationhood based on ethnic and cultural identity. More than twenty years ago (1979), I visited Singapore with my wife and son for the first time. Singapore is a city-state established after World War II. It is composed of a diversity of ethnic peoples. The majority are Chinese (70%) while the others are Malays (20%) and Indians (10%). As soon as we got off the airplane, we realized we were in an unusual Asian country. Singapore is clean, law-abiding, prosperous and safe. It is like the Switzerland of Asia. As they were fetching our luggage, we were addressed by a porter in perfect Chinese who said, "Can I help you with your luggage?" I was very surprised and said, "How come your Chinese is so good?" His answer was, "Of course, I am Chinese". This response expressed the pride of self-identity. Here was somebody who probably had never been to China, in a place that had nothing to do with the Chinese state, and yet was in a community which identified itself as Chinese. Later on I learned that the Chinese in Singapore, like many Chinese everywhere, promoted their own education by the teaching of the Chinese language, and in that process they maintained their identity with Chinese culture.

From the point of view of ethnic origin, language and culture, there is no doubt that Taiwan is Chinese. I know from my personal experience during World War II that China is a great and heterogeneous country. I traveled in Guandong, Sichuan, Hunan, Hubei, Yunan, Shanghai, and Beijing. Each of these places has its own dialect, and to some extent is different in its customs and practices. Some of them also had different historical experiences. Still there is no doubt they are all part of China. Some people in Taiwan make a great deal of the differences in their historical experience. Since the emigration of the majority of Taiwanese from the mainland province of Fukien more than two hundred years ago, Taiwan underwent first the occupation by the Dutch, and then for almost fifty years the occupation by Japan. Still, for most of the time, Taiwan was ruled by the Chinese of the Chin Dynasty. Ex-President Lee looks back fondly on the experience of Japanese occupation. Lee considers a part of his own cultural heritage to be Japanese. But the experience under the Japanese was short and superficial. Consider the fact that President Chen has had no contact with Japanese culture. Japan is as foreign to him as it is to me. Then some Taiwanese make a great deal of the events of "February 28". That is the day, in 1950, after the Nationalist regime had newly arrived in Taiwan, and brutally suppressed what they considered was an uprising of the people in Taiwan. One must consider the fact that the Nationalist regime was brutal and oppressive in many ways in its final days. Many people on the mainland suffered under them before they left. The actions of the Nationalist political authority should not be identified with the Chinese nation. The events of February 28 should not be ascribed to all Chinese. My personal experience in Taiwan leads me to the conclusion that Taiwan is as Chinese as any other province on the mainland. One of the joys of my being in Taiwan was my immersion in Chinese society and culture. I had the experience of speaking Chinese, writing Chinese, reading Chinese books and periodicals, making Chinese friends, and eating Chinese food. What I experienced was the China that I had missed in the United States for over fifty years.

From the narrow point of political authority, Taiwan can become independent as proposed by the platform of the People's Progressive Party of President Chen. She can separate herself completely from the Chinese state across the straits. This state can give itself a name that does not even include the word "Chinese". None of these changes can alter the fact that such a state would still be Chinese in terms of its nationhood. The Taiwanese cannot change the fact that they are ethnically and racially Chinese and that their culture is totally Chinese. They can choose to avoid Mandarin, the universal spoken Chinese language, and introduce the use of "Taiwanese". But this cannot alter the fact that Taiwanese (or more accurately a southern Fukien dialect) is a Chinese dialect like the many other Chinese dialects. (Interestingly, making the southern Fukien dialect the official language would offend 20% of the population, who speak Hakka.) The Chinese in Singapore wisely teach Mandarin, and avoid their own dialects that are frequently incomprehensible to other Chinese in Singapore who speak different dialects. What the people in Taiwan can choose is a different state. What they cannot choose is a different nation.

For most Chinese, the century after the 1840s was defined by oppression and shame imposed by the Western and Japanese imperialists. Only after the victory of World War II was this situation finally reversed. To most Chinese, irrespective of political beliefs, the return of Hong Kong and Taiwan to China is restitution of justice. Vice Premier Qian Qichen of mainland China declared on August 24, 2000 a new formulation of the "One China" principle. "There is only one China in the world, mainland China and Taiwan both belong to this China. Its territories are inviolable and cannot be separated". This version eliminates the objection in Taiwan that Taiwan should return to Communist China. It seems to me that Chinese of all different political affiliations can support this principle.

The Taiwan Straits and the United States

The relationship of the United States to the Taiwan problem is complex. On the one hand, the United States is bound to respect the Cairo Declaration of 1943 that is the basis of the peace with Japan after World War II in which Taiwan was clearly returned to China. On the other hand, there is a strong vein of opinion in the United States that wishes to maintain its hold in East Asia. Since World War II, Taiwan has been called the largest unsinkable aircraft carrier of the United States. This is an important consideration, since some people in the United States consider China to be a potential adversary of the United States. Some of them may favor an independent Taiwan. Taiwan is fully aware that without full support of the United States, it would not be possible to achieve independence in Taiwan. It behooves the independence people in Taiwan to consider the implications of survival if it were only possible by support of the United States, even if it is the sole superpower. We already have one state that exists under such conditions. That state is Israel. And despite the support of the United States, Israel has not known peace since it was founded.

Conclusion

I stated my support for the concept of "One China" in the above article written in 2000. What I meant was that there should be one Chinese "state", because in terms of nationhood, Taiwan is already part of China. However, I did not explain or discuss how this one Chinese state could come about. Mainland China has not given a time limit, although it has stated it would not wait indefinitely to achieve one statehood. I realize that once negotiations are resumed, innumerable problems will arise to prevent reunification. Many of the problems seem at present insoluble. Many people in Taiwan assert that if only mainland China would become democratic, then the possibility of reunification would be more attractive. I confess I do not know exactly how such problems can be resolved. But experience of the last twenty years teaches us that many things can change in ten or twenty years. Twenty years ago, I listened to a Taiwan diplomat in the United States, complaining how poor mainland China was, and how economically prosperous Taiwan had become. Now we note that practically every one admits to the vast strides made by mainland China economically, and paradoxically, most people in Taiwan believe that Taiwan's economic future depends on a close relationship with mainland China. Right now there are hundreds of thousands of Taiwanese doing business on the mainland even though its government does not overtly support it. While the mainland is still far from being a democracy, it has become far different from the totalitarian state that it was in the 1970s. It is apparent that both Taiwan and the mainland have made vast strides in its economy in the last thirty years. In the beginning, Taiwan was far ahead, but China has had a steady growth for the last thirty years, which some have called the new economic miracle. I believe that the economic miracles in both countries represent the fruits of two factors. One is the heavy emphasis on education in both countries, which can be considered a derivative of the Confucian ethic. In the Chin Dynasty, Confucian ethics was a factor in China's backwardness, because it limited education narrowly to the classics, without regard for advances in the sciences and modern technology developed in the West. The difference now is that in both mainland China and Taiwan, education has been completely modernized, and is comparable to the educational systems in the advanced nations in the West and Japan. But in addition, the old Confucian ethic that motivates children and parents to strive for the highest goals in education has remained operative. The second factor is the intelligence, and work ethic of the Chinese people. Both factors I believe are integral components of Chinese culture. They have been able to contribute to prosperity on both sides of the Taiwan Straits because of the peace that has prevailed in the last half century.

I have observed in my short time in Taiwan evidence of much improvement in what I had previously considered were problems. For example, with arrival of an ultra modern subway system in Taipei, and a rigorous enforcement of rules of common courtesy, attitudes of subway riders have changed. Previously, in Taiwan as on the mainland, getting into a bus or a train was a hassle. People would disregard the line and try to squeeze in. Now passengers stand in line and enter and exit in an orderly fashion. What I was most surprised to see was that very often I, as an elderly, was offered a seat. My experience is that now this happens more often in Taiwan than in the United States. During recent visits to China in 2001 and 2002, I also observed the rapid improvement in what I call the infrastructure. In the public or national parks at Huang-Shan and Hang-Zhou, there was uniformly strict observation of rules concerning disposal of trash. Crowds moved about in an orderly rather than a menacing manner. Equally surprising was the improved quality of lavatories in public places.

Corruption has been a perennial problem in China. It is prevalent in China and to a lesser extent in Taiwan. It too may be considered part of Chinese culture that must be changed. I am even optimistic in this regard. There are places where Chinese are in charge, such as in Hong Kong and in Singapore, where corruption has all but disappeared. I am optimistic that in time, Taiwan and the mainland will approximate each other economically, socially and even politically, such that reunification, like their other common problems, will be resolved naturally.

| 18 |

TRADITIONAL CHINESE MEDICINE

C hinese medicine is an integral part of Chinese culture[110]. It has been practiced in China and in Chinese communities for at least three thousand years. The oldest classic of Chinese medicine, the "Canon of Medicine of the Yellow Emperor" is ascribed to the mythical "Yellow Emperor", who presumably lived around 2698–2598 B.C. The text was compiled during the Spring and Autumn and the Warring States Periods (770–221 B.C.), a period of history during which original thinking in China reached its highest peak. The "Canon of Medicine" already established how Chinese thought about disease, bodily function and vital forces.

During the Spring and Autumn Period, Laotse and Chuantse elaborated the nature of "dao", or the "natural way". The driving forces of nature were divided into "yin" and "yang" types. Yin and yang are opposites; they are the plus and the minus, male and female, sun and moon, etc. Their interactions make for harmony in the universe. The "Canon of Medicine" used this idea to explain health and disease. Health is the harmony of yin and yang forces, and disease is disharmony.

These forces reside in the internal organs of the human being, or the system of "zang" and "fu". Diseases are categorized as deficiencies of organ function, such as deficiencies of "liver" or "spleen" functions. There are five solid organs (zang) that are yin in nature (lungs, heart, liver, spleen and kidneys), and six hollow organs (fu) that are yang in nature (gall bladder, stomach, small intestine, large intestine, urinary bladder, and an anatomically unknown organ called "san jiao" (three body cavities)). These organs interact and communicate with each other through an elaborate system of channels throughout the body, which are precisely described, but for which we have as yet no anatomic evidence. An invisible "qi" circulates through these channels.

Qi, or chi, literally means vapor or gas. It may be thought of a type of energy, or electricity, an invisible, but powerful force. But unlike electricity, it may have a strong psychological component which is as yet undefined. The whole system of acupuncture is based on regulating qi in the described channels on the surface of the body. Qi is also the basis of a whole set of popular exercises, called "qi-gong", or the "qi exercises", which like yoga, can produce remarkable acrobatic effects in the initiated. So there may be something substantial corresponding to qi.

Diseases are specific dysfunction of the yin and yang forces acting in different organs. Dysfunction may be due to external or internal causes. The six external pathogenic factors are wind, heat, cold, dryness, dampness and summer heat. During the East Han Dynasty (22–220), Zhang Zhongjing, an eminent medical scholar, wrote a classic, "Treatise on Febrile Diseases caused by Cold" (111). "Cold" is one of the pathogenic factors mentioned above. This treatise may be considered the infectious disease classic of Chinese medicine. Based on concepts of the "Canon of Medicine", he describes in detail febrile illness syndromes based on clinical observations. He explains their origin and diagnosis, using the pathophysiological concepts described above.

Another Chinese medical classic is Li Shizhen's "Compendium of Materia Medica" of 1590. It is the monumental achievement of a scholar of the Ming Dynasty (1518–1593). It is an encyclopedic glossary of medicinal plants containing botanical and pharmacological observations.

While traditional Chinese medicine was always practiced where there were Chinese, its acceptance in recent years has had its ups and downs. It was largely debunked during the nineteenth and first part of the twentieth centuries, along with other concepts of traditional Chinese culture by Westerners and Western-trained Chinese intellectuals. It received its modern revival under Communism. Mao Zedong himself encouraged the practice and study of traditional Chinese medicine in China in the 1950s and 1960s. As a matter of fact, Western and traditional Chinese medicine received equal recognition in the Chinese constitution! The Communists instituted the system of combining both types of medicines in their hospitals, and in providing health care. In practice, they have relied on Western medicine in public health, and in the diagnosis of diseases. Traditional Chinese medicine was resorted to when the patient so chose, or when no Western therapy was available, or when it failed. The renewed popularity of traditional Chinese medicine extends to Taiwan and other oversees Chinese communities. There are schools of traditional Chinese medicine on the mainland and in Taiwan. In Taiwan, students often get degrees in both types of medicine, in order to appeal to their patients.

"Alternative medicine", including traditional Chinese medicine, has also become popular in the West in recent years. Therefore, it is important for China and Taiwan, as well as the West, to understand how traditional Chinese medicine can be made scientific so that it can become useful.

My view of traditional Chinese medicine is that one should find out scientifically what is valuable in it. But first we must recognize what it is not. It is not scientific medicine. Great harm is done by sharply separating traditional Chinese medicine from so called Western scientific medicine and assuming that they are equal in validity. Western scientific medicine is international in nature, and does not belong only to the West or the East. Scientific medicine should eventually encompass all medicine, because all medicine should be scientific. Traditional Chinese medicine belongs to a great ethnic culture, but it is only one of the many ethnic schools of medicine, none of which is scientific as we understand it. There were schools of alternative medicine in the history of medicine in the West. We associate them with the names like Hippocrates and Galen. There are ethnic schools of medicine among practically all developed, and less developed cultures of the world. The nature of these ethnic schools is that they are based on dogma, either written or oral. They are not observationally, or empirically based. They are not structured in a way so that mistakes can be eliminated, and new knowledge may be added. One Chinese book I read compares the doctrine of yin and yang in Chinese medicine to Newton and Einstein's laws of physics. This is a great misapprehension. In science, Newton's physical laws of the universe have been disproved and superseded by Einstein's. This was orderly progress of science. Yin and yang may be insightful, they may be holistic, they may be poetic, but they are not part of a scientific hypothesis, because there is no way to disprove them.

There can be no scientific discourse about traditional Chinese medicine at the theoretical level. The Chinese theories of anatomy, physiology and pathogenesis of diseases are self contained, and exclusive of outside influence. There is no way to change them in the light of new observations. Take anatomy. It doesn't help to tell Chinese physicians that the kidney has no sexual function, because anatomic and physiological observations do not reveal any. The Chinese dogma is, however, that the kidneys are the center of sexual function, and all the yin and yang forces associated with it.

They would rather maintain their dogma about the functions of the kidney rather than change their idea of what an organ is. Admittedly, one has to be careful in making such dogmatic statements. In the eyes of traditional Chinese medicine, the kidneys also have a role in hemopoiesis (formation of blood), since it is said that marrow is in the kidneys. Fifty years ago I would have said that they are dead wrong. Around that time, a colleague of mine, Allan Erslev, at Thorndike Memorial Laboratory at Harvard where I was a fellow, discovered the erythropoietic (red blood cell growth) factor[112]. Of all places, it was made in the kidney! So the Chinese may be said to be at least partially right. But this actually proves what being scientific is, there is constant progress, and one can be proven wrong. How can we learn from traditional Chinese medicine, and what can we do to make it scientific? I believe there is an answer.

What Is Meant By "Being Scientific?"

Even though we are in an age of recurring scientific discoveries, it is remarkable how unscientific the thinking of the common person is. This point has been made by the popular American astronomer Carl Sagan, who pointed out that superstitions, misconceptions, and lack of understanding about science abound in daily life[114]. They prevent rational and scientific decision making. This is true in the United States, but it is also true even among the well educated in China and in Taiwan.

Perhaps the most important point about being scientific is to understand the nature of scientific evidence. Such understanding does not come about by our learning about or how to use the latest technological gadgets. It comes about by understanding how scientific knowledge is accumulated. In this regard, I was heavily influenced by Karl Popper during my college days at Harvard. He provided me with a clear understanding of what being scientific is. Everyone knows that modern science is based on empirical evidence. Popper defined clearly what scientific evidence can and cannot do. For example, the proposition "all crows are black" cannot really be proven. But it can stand as a hypothesis until empirically a crow is found that is not black. To Popper, science is a series of propositions or hypotheses that are falsifiable by empirical evidence.

In more recent years, Popper's critique has been effectively applied to medicine (see Chapter 5).

A similar critique of the pathogenic theories of traditional Chinese medicine would also find them to be unscientific. This would be also true of the propositions and tenets of ancient western medicine from Hippocrates to Galen. Most of these unproven doctrines have been given up. Generally speaking, what can be made scientific in traditional Chinese medicine is its therapy. Classically, the therapeutic modalities of Chinese medicine consist of acupuncture, moxibustion, and plant concoctions. In order to utilize these therapeutic modalities, it is not necessary to understand or agree with their theoretical, pathogenic justification. It is sufficient to prove them therapeutically effective by rigorous randomized controlled trials (RCT's). If proven, therapies can then be accepted in the armamentarium of therapy that is common to all human medicine.

How to Make Traditional Chinese Medicine Scientific

There has been a revolution in the evidence required for therapy in the last fifty years. When I was a medical student, we were told that medicine is "a combination of art and science". Much of the therapy was in the area of art, largely unproven and anecdotal. It is only after World War II that one by one, all new therapies practiced by western medicine, and frequently ancient medicaments, like digitalis and quinidine, have been scrutinized by RCT to become evidence-based. The inevitable conclusion of this revolutionary experience is that all therapeutic modalities, whether Chinese or Western, must eventually be proven to be evidence based in order to be acceptable.

An English statistician, Austin Bradford-Hill (1897–1991), was responsible for introducing RCT. He did not get the Nobel Prize, but he should have gotten one. One of the first RCT's was a trial sponsored by the British Medical Council in 1948, of streptomycin in the treatment of tuberculosis[115]. He took a group of patients and randomized them into an experimental group that received the therapy and a control group that did not. The outcome in each group, whether cure or increased survival, was recorded and analyzed statistically. Efficacy was based on a statistical difference between the two groups. He was also among the first who discovered the causative role of smoking in cancer of the lung[116].

His methods now seem obvious and are acceptable to all, but they have taken the last fifty years to take hold. The main reason is that the prerequisites for proceeding with an RCT are not always easily met. The most important one is that those in charge of an RCT must accept the premise that, whether the therapy is or not effective, is unknown. Only then can one ethically justify an untreated control arm in the RCT. Practitioners convinced of the efficacy of their treatment cannot easily accept such untreated controls. Even in the United States it is difficult for some surgeons to accept RCT if effectiveness of their surgery is questioned. For a practitioner of traditional Chinese medicine imbued by its doctrines, and convinced of its efficacy, a rigorous RCT would not be lightly accepted. In order for them to accept RCT, they must be prepared by a vigorous course of education, just like western physicians. The mainstay of such education is presentation of the voluminous documentation that so much of non-evidence based therapy is either useless or harmful.

My interest in traditional Chinese medicine began in the 1980s and 1990s when I was active in conducting clinical trials in antivirals. I realized that scientifically conducted clinical trials had revolutionized western medicine since World War II. It is the basis of "evidence-based medicine", which is more or less synonymous with "scientific medicine". It occurred to me that, while the pathogenic theories of diseases in traditional Chinese medicine are largely inscrutable to western scientists, and cannot be made "scientific", the therapeutic modalities of traditional Chinese medicine could be tested by randomized control trials (RCT) and be made scientific. This insight led me to sponsor, with the help of President Wu Cheng-wen of NHRI, a "Symposium on Clinical Trials in Traditional Chinese Medicine" in Taiwan in 1996. Participating in this conference was Dr. Hsieh Gui-hsiung, one of the few western physicians in Taiwan who had actually conducted a RCT for asthma using Chinese herbs[113]. There were also experts in traditional Chinese medicine, and others who were interested in evidence-based medicine in attendance. This conference popularized the idea that RCT's were necessary. When I arrived in Taiwan in 1997, one of my objectives was to further promote the idea of RCT in traditional Chinese medicine. In the intervening years, a consensus has developed in Taiwan that RCT's are necessary for Chinese medicine. Even though I did not have time to conduct RTC's, as I was fully occupied by the work on antibiotic resistance, I was still interested in it and wrote several articles and gave a number of talks on it. I realized that good ideas are often superficially accepted, but they may not be really understood. This is the status of RCT in traditional Chinese medicine in China and in Taiwan today.

One of the simplest, and most popular methods of extracting therapeutic benefits from Chinese medicine, has been to purify chemically its potentially beneficial ingredients. More than sixty years ago, K. K. Chen isolated ephedrine from the Chinese herb "mahuang", known to Chinese practitioners to be effective against bronchial asthma[117]. More recently, chemists on the mainland have successfully isolated artemisin from "ching hao su", another Chinese herbal medicine. It is now one of the mainstays in the treatment of falciparum malaria, a disease that is increasingly difficult to prevent and treat, because of mounting drug resistance. Such accomplishments, although dramatic, have actually been remarkably few in number, considering the vast infrastructure of the pharmaceutical industry and research institutions looking for such discoveries in Chinese medicines, not only in mainland China, Hong Kong and Taiwan, but also in Japan. Japan's interest in Chinese medicine is almost as ancient as China's. One reason why not more pure drugs are extracted from Chinese medicines may be because the Chinese practitioner rarely uses pure chemical compounds or single herbs. They use concoctions of herbs, whose precise constitution often eludes precise chemical characterization.

The testing of pure compounds derived from herbal medicines is actually not directly relevant to providing evidence for the practice of Chinese medicine. The pure compound is not what the Chinese practitioner uses. These compounds are used in patients diagnosed by western medicine, like asthma or malaria. Ancient China was neither aware of infections by microorganisms, nor of conditions such as malignancies. "Mahuang" and "chin hao su" are used by Chinese practitioners for diagnoses made according to Chinese theories of pathogenesis. Diseases, according to the Chinese practitioner, come from disturbances of yin and yang, or from environmental assaults, such as "cold" or "heat". These disease concepts may be incomprehensible according to western medicine, but whether or not the medicine works in this or that disease condition according to Chinese theories of pathogenesis can be verified by RCT. RCT requires neither chemical purity, nor correct anatomic and pathological diagnoses.

What I am proposing is a systematic, comprehensive application of RCT to all the major modalities of therapy of traditional Chinese medicine as used by its practitioners. Traditional practitioners can determine how to do this systematically so that most important treatment modalities are tested. One need not have pure chemical compounds to do RCT's. Impure concoctions are permissible, but they must at least be standardized so that experiments can be repeated.

In order to test a Chinese prescription fairly, it must be used for diseases diagnosed by Chinese practitioners. These include diseases of disturbances of yin and yang, deficiencies of "organ functions", external forces of pathogenesis. There are no terms for diseases such as cancer, infectious diseases, or "Alzheimer's disease". Therefore in a proper RCT, the traditional practitioners must be involved with every phase of the trial, but especially in the preliminary choice of diagnostic categories that define the treatment groups. An RCT conducted under these terms would then be meaningful for the Chinese practitioner, and not just for the western practitioner.

I believe the time is right for such an approach, both in the mainland and in Taiwan. There are in Taiwan many traditional practitioners now who have been trained in both traditional Chinese medicine and in western medicine. They are particularly suitable for undertaking RCT's such as I suggest.

Unless we have a cadre of traditional practitioners who understand and believe in the importance of RCT for the betterment and survival of traditional Chinese medicine, and who are involved in carrying them out, traditional Chinese medicine will not escape the trap of remaining "unscientific", and eventually being eliminated.

Misconceptions about Traditional Chinese Medicine

There are certain misconceptions about traditional Chinese medicine in Taiwan as well as in mainland China.

I have already referred to the first misconception, which is to consider traditional Chinese medicine and scientific medicine as separate but equal. Eventually, all types of medicine should be scientific and international. In the meantime, one must accept the fact that Chinese medicine is not yet part of scientific medicine.

The second misconception is that Western medicine is good for acute diseases while Chinese medicine is better for chronic diseases. While it is certainly true that only Western medicine can treat certain acute diseases such as bacterial infectious diseases, and diseases whose therapy requires surgery, there is no evidence that generally speaking, Chinese medicine is better for chronic diseases. To cite but one example, without

Western medicines such as insulin, it would be difficult to maintain the health of patients with a chronic disease such as diabetes, especially of the Type 1 type.

The third misconception is that Chinese medicine is "restorative", and better for rehabilitation. Chinese medicines are said to be best for improving the "immune system." This concept is prevalent among many modern Chinese. It is difficult to understand what they mean by the "immune system." The fact is, I know of no Chinese medicine that improves any part of what we understand as the immune system, if we classify it rigorously as humoral or cellular immunity. This concept is used to support the frequent use of Chinese medicines for diseases for which Western medicine has no specific cure. This includes especially the degenerative diseases, such as rheumatic and osteoarthritis, degenerative diseases of the spine, and diseases associated with increasing age. Admittedly steroids, anti-inflammatory agents, and analgesics are commonly used for these conditions in western medicine, and they are not curative, but which Chinese medicine is better? There is none that I know of. This is where RCT is urgently needed, to discover situations where Chinese therapies may be superior.

The fourth damaging misconception is that Chinese medicines are non-toxic, while Western drugs can be extremely toxic. This basic misconception has led the U.S. FDA to consider herbal medicines as nutritional supplements, rather than drugs. Instances of toxicity of herbal medicines have recently become commonplace. A well-known example is the toxicity of a common Chinese medicine called "fang chi". As reported in the *New England Journal of Medicine* in 2002, this drug has the unusual property of producing carcinoma of the ureter[118]. Just because Chinese medicines are mostly derived from plants cannot guarantee their lack of toxicity. The fact is most of our drugs come from plants, before they are chemically synthesized. Chinese medicines must be tested for toxicity just like any other drug. That the discovery of toxicities of Chinese medicines are reported in the West rather than Asian countries is evidence that inadequate attention is paid to the toxicity of Chinese medicines in Asian countries, where they are most frequently used. Only recently in mainland China, and in Taiwan, has there been the beginnings of a system for toxicity surveillance as developed in the United States.

Traditional Chinese medicine is an ancient system of medicine, which has stood still for thousands of years. In the meantime, medical science has advanced in the West to the point where a therapy must be evidence-based before it can be accepted. This new requirement must apply to all therapies, even those enshrined by antiquity. Experience tells us that many therapies that cannot pass the test of RCT must eventually be abandoned. All modalities of Chinese medicine must be proven by RCT if they are to become part of the therapeutic armamentarium of modern medicine.

chapter

| 19 |

RELIGIONS AND SPIRITUALITY

There are questions about life which concern every one. What is its purpose? What is its meaning? Where do I come from? Where will I go? These questions are in the realm of religions and spirituality.

Religion, like sex, is said to be a private matter. One is not expected to talk about it. This has always seemed a bit strange to me. Religion is not like sex or politics, where personal preferences and tastes are respected. These are legitimately considered to be in the realm of "privacy". The tenets of religions are very much a problem of knowledge, which may either be true or false, or they may be partly true and partly false. In any case, like any question about knowledge, their elucidation should be an object of common human concern. Disputation and discussion should be encouraged rather than suppressed. One's views should be of interest and concern to others. Instead they are hidden and often shunned.

This is the case not only among strangers, or among friends, it is also the case within the family; between siblings, between parents and children, and even between husbands and wives.

One person close to me throughout my life was my father. He was baptized and brought up as a child as a Protestant (Lutheran) in Hunan, China. His illiterate mother was a devout Christian. He went to missionary schools. That does not mean he was religious. He was like so many nominal Christians, particularly during his adult life; neither formal nor informal religion seemed very important to him. While he was meticulous about instructing me in many matters as a child, I do not remember a single occasion when he talked to me about Jesus or Christianity. It was after he retired and after he was over seventy years old that he became interested in the

288

church and religion again. He became a founding member of the Chinese Lutheran Church in his neighborhood in San Francisco. He began to read the Bible in Chinese faithfully, almost everyday. We had a discussion about Christianity for the first time when both of us were fairly old men.

My experience with Christianity has also had its ups and downs. I too was baptized and confirmed as a young boy in Brooklyn, N. Y. There were times in college when I toyed with fundamentalist Protestantism. I remember a serious young graduate student at Stanford who befriended me. He tried to convert me. One day while walking together, he suddenly stopped and began to pray fervently for my conversion, and my "acceptance of Christ". Instead of being converted, I was deeply embarrassed by this experience. Still, while I was a graduate student and medical student at Stanford University, I was a member of a small Christian group which met periodically. Ten years later, Carol and I were members of similar young couples Christian group attached to Calvary Episcopal Church during our early days in Pittsburgh. But try as I might, I never reached the point where I could call myself a true "believer". The Christian dogma, epitomized in the "Apostle's Creed", seems to stick in my throat whenever I recite it in church. I also had problems with the doctrine of "Trinity", that God exists in the form of the Father, the Son and Holy Spirit. I could not understand the idea that Jesus, Son of God, was God. The idea only makes sense to me insofar as all men are sons of God, but then we cannot all be gods. On the other hand, I developed an interest in and admiration for Jesus the man. This includes an admiration and appreciation for his profound belief in God, as he understood him. In general my understanding of Jesus derives from the writings of the New Testament, or the four Gospels. I admire the service that Martin Luther rendered Christianity during the Reformation, when he translated the Bible to vernacular German, and opened these remarkable documents to the common man.

Even though I cannot share Jesus' absolute belief in an omnipotent God, who is also loving like a father figure, I realize that the attractiveness of Christianity comes from this faith, irrational or mystical it may be. "Love your neighbor as yourself" is an emanation of God's love for man, or so it is said. Another expression of the all importance of love in Christianity is the simple statement that "God is Love". These words were indelibly inscribed in my mind when I saw them in three Chinese characters hanging on the wall in the home of an American medical missionary in China when I was at Harvard College. In its emphasis of "love", Christianity is unique among all religions. Basically I believe that great religions enshrine and sublimate the important, basic and worthwhile human characteristics. Christianity has taken a basic and noble human emotion, "love", and sublimated and enshrined it as a deity. There seems to be no worthier human sentiment to do this to. Because of that insight alone, Christianity is to me a worthwhile religion.

My understanding and appreciation of Christianity is largely secular. When I tried to explain to my father my views of Christianity, I confronted a stone wall. He was not interested in my views nor would he discuss them with me. To him, "believing" was basic to his religion. Believing meant believing without reservation those basic dogmas I have found so difficult to accept. However not only did I, an avowed agnostic, find them difficult to believe, I discovered that he had similar difficulties. He struggled valiantly against them. I felt that in his eagerness to "believe", he lost appreciation of the attractiveness of Jesus the man. I never heard him express admiration for Jesus' human qualities. I shall never forget overhearing his prayers during the last days of his terminal illness. He was supplicating, lamenting, deploring and imploring an unwilling Deity to save his life. It was pitiful. It was almost degrading. I felt a true Christian should know how to die. I asked his minister, Pastor Kuo, to talk with him about death and to comfort him. He did not do so. Nor did I have the courage to do so.

I believe that the inability of so called Christians to truly believe is not an exception but the rule. For that matter, it may be the same for all religions. Even in medieval days when faith was taken so much more seriously, and often for granted, we have the memorable prayer of the great St. Augustine, "O God, help my unbelief!" Here is the dilemma of the believer. On the one hand, it may make him a hypocrite. And religions turn out a lot of them. Even the Bible tells us so. At the other extreme, so-called believers can become fanatics. Faith is used to justify many actions, including violent ones. Religions condone, absolve and permit bad as well as good deeds. Blind faith is the source of much of the bigotry, intolerance, injustice and violence of organized religion. Wars are very much part of the history of religion. This is certainly true of the three great monotheistic religions, Christianity, the Moslem religion and the Jewish religion. Buddhism may be an interesting exception.

My experience with Christianity, including my understanding of my father's experience with it leads me to a much broader definition of religion as a problem for man. To me, it no longer is a question whether or not to believe in a religion. I simply have been unable to accept the basic religious dogmas. I do not try to any more. Rather I am willing to open my mind and heart to different religious ideas, creeds and practices and look at them in a sympathetic light. I am very much influenced in my attitude by Huston Smith, who uses this approach with the major religions of the world[119]. Ken Wilber does this in a more holistic and scholastic way. In his many books, he pursues the singular grand synthesis; of science and spirituality, of the ancient and modern, and of the East and the West[120]. I approach religion very much like other problems I have met in my lifetime. They have to be studied, understood and appreciated. Understanding religions and spirituality is an educational process; a valuable part of life's experiences.

Part of this education is knowing great people. My knowing Jesus through the gospels is this type of education. But great people are also living among us today. It is enlightening to know and experience such a person at first hand.

Bettie: A Great Woman

Bettie Chu is the wife of Charles, who was one of the first friends I made in 1947 when I arrived at Harvard. He was a Chinese graduate student in political science. Bettie was born Bettie Simmons Wilson, born and raised

Bettie and Charles Ji-yun Chu at Connecticut College, New London, Connecticut, 2003.

in California. They met when both of them were students at the University of California, Berkeley. Shortly after Charles and I met at Harvard Yard, he took me to Arlington, a suburb of Boston, to meet his newly wed. That was when I began to know and appreciate this remarkable woman. And to enjoy knowing her for more than fifty-five years I count as a rare privilege and blessing for me.

Bettie is so much like and yet so different from Charles. Both have this rare gift for easy friendship and intimacy. Both are informal and wonderful hosts. Both entertain and offer hospitality seemingly without effort. That is why their house is always full of people. Not just guests who drop in and chat, but guests who stay for lunch and dinner, and who stay as house guests for days, weeks, even months. A few have even stayed for years. Every time we want to visit them, it is always "come any time". But I know better than to assume that to be true. Going to their house is like making a reservation at a popular inn during the height of the tourist season. Very often I discover when we want to see them that their house is booked days or even weeks ahead of time. For some of us like my wife Carol and me, who do not have many guests, let alone house guests, this is nothing short of phenomenal.

Bettie projects something unique. What she projects and emanates is love. Love for those around her and interact with her. After associating with her, one realizes that surprisingly, she can love the whole world. For most people, love of one's neighbor is an abstract notion, an ideal to be striven for, something to be indulged in during impulsive moments, a religious commandment but a practical impossibility as a life style. For her, it is what she manifests in everyday life. It is in the air around her; it is her. She shows love when she deals with you, whether you are a member of her family or a friend, but most remarkably, she shows it to you if you are an acquaintance or even just a stranger.

One day in 1994 or 1995, a stranger did come to her and told her he was in need of a thousand dollars. He said he was a student who had to travel and was in need of funds. Since it was a holiday, she took him to a money machine, took the money out of her bank account and gave it to him, no questions asked. There was no discussion of how or whether the money would be returned. To us mere mortals, including Charles, her husband, this was inconceivable behavior. He was beside himself with frustration. Thinking about it, to me, this is the behavior of either a fool or maybe a very wealthy person. Or it could be the behavior of a true saint. Bettie is no fool. And she is not wealthy.

What makes Bettie unique is that she is also eminently practical and wise. She has solutions for daily, personal problems. That's why she is valued as a confidante and advisor. Our daughter Bettie (named after Bettie, of course) had her share of problems growing up during adolescence. She would be on the phone for hours talking to her "Gan-ma" (god-mother). She would drive from Pittsburgh, Pennsylvania to New London, Connecticut and back on the spur of the moment to seek her advice; and seek renewal. Bettie always gave sensible, practical advice. Even though the advice may not be carried out by the recipient, whatever she offered seemed to be suffused with love or it is transfigured by love. Love for the one being advised, and for ones, frequently unknown to her, which the advice might involve. Her advice is feasible, practical and eminently reasonable. So besides love, she emanates wisdom.

From early on, I regarded Bettie as my teacher. I wanted to emulate her ways in dealing with people, including children. I used to watch Bettie when she was bringing up her own children. Eating always seems to be a problem with children. They do not eat what you feed them. Or they make a mess of their food. Eating often ends up being a battle ground between parent and child. It was never that way with Bettie and her children. She would put a portion of food in front of the child at regular meal times, with never a word of encouragement, discouragement or coaxing. If the child made a mess, she just cleaned it up and nothing was ever said. Leftovers were simply taken away without comment. All meals passed in this peaceful, wholesome atmosphere. I do not think any of her four children ever had a problem with eating. She demonstrated to me that problems of eating are usually not problems of the child but of the parents. I remember trying to practice what I learned about this on our children, Bettie and John, when they were small. One day we were having corn on the cob during the height of summer. In order to solve the problem of preventing them from making a mess, but not saying a word of reproach, I dressed them in their bathing suits, put them in a bathtub, and gave them their corn on the cob. They could be as messy as they wanted and we did not have to be our usually compulsive selves. They still remember this with great enjoyment. They will always enjoy corn on the cob!

In raising her children, I noticed that Bettie never raised her voice or scolded them. I never caught her "lecturing" them, but she would spend hours communicating with them and reading to them. She never used the "you" word. She is never accusatory, with either children or adults. Her children were taught by example. Most behavioral problems were prevented before they occurred. The proof of her approach were the results. They were beautiful to behold. All of her children were natural, well behaved, gentle and considerate, like herself. Her methods seemed so simple, and Bettie herself carried them off without apparent effort. But they in fact are very difficult. I think they are particularly difficult for some one with a Chinese background. In China, there is a great deal of emphasis on active instruction of children and minors. I don't believe I was ever able to copy Bettie entirely. But I did come to realize the limitations of active preaching and the shortcomings of physical punishment as a pedagogical device.

Bettie deals with adults and the adult world in the same spirit as she treats her children and family. Everything she touches is clothed in love. She is totally devoid of malice, rancor, envy, jealousy or any trace of pettiness. But she also has a great deal of technique. She is extraordinarily perceptive; she is the epitome of tact and consideration. She listens to, understands and appreciates apparently contradictory view points. Her whole approach to human relations seems effortless and seamless. She has the touch of both a saint and a genius. That is why it is so difficult to copy her. One either lacks love or wisdom, rarely does one possess enough of both. Over the years, one of the most favorite questions asked by me and within our family when confronted by a particularly difficult human problem, is "How would Bettie handle this?" I even find myself asking this question when dealing with problems in my professional life. Even if I cannot solve them entirely her way, I try to emulate her. Thinking about her ways usually makes me reach for something kinder and more considerate.

In more recent years, I have also come to regard Bettie as my spiritual mentor. Again she may know nothing about this. Like in other things, I learn about her approach to the spiritual life by talking with her, by watching her behavior and the way she lives her life. Observing what she reads makes a big impression on me and makes me think. She is constantly studying and reading in this area. I too am thinking and learning about these things. Just talking with her and being around her encourages me in my efforts. This is what I love doing during the private retreats we have been having in the last few years; just the four of us in some isolated vacation spot for a week or so. I feel I am following her in a common quest, although we do not verbalize this much. Recently she was surprised to learn that I had such interests, thinking I am perhaps an intellectualized being surfeited with rationality. I was pleased that she recognized the emotional and spiritual side of me.

Having known Bettie has made me rethink the definition of greatness. What makes for greatness in a human being? Fame, wealth, power, beauty, influence? Opinions vary and definitions vary. Most definitions seem to have one common denominator. That is fame. Great men are famous for one reason or another. But Bettie has me also questioning this definition. Even though Bettie is not famous, and she may never be, for she is known only among family and friends, I consider her a great woman. Bettie's greatness lies in her ability to express love in daily life among people she comes in contact with. This ability touches all those around her. She is a luminous personality who shines with innate greatness. Her greatness is inspiring, nurturing and invites admiration and emulation.

Knowing Bettie has been an education for me. She teaches love and goodness by example. This is better than any book or sermon. She is not a formally religious person, although she attends Friends' meetings occasionally. She is very much interested in the spiritual, in meditation and silence.

Where does her goodness and love come from? I don't really know. I cannot say it comes from her religion. I cannot even say that it comes from her spirituality, even though I believe she is a deeply spiritual person. Her goodness and love seem to be just part of her. If there is a God, there is more of him in her than in most people.

The Religious Quest

There was a Chinese Protestant minister in Pittsburgh, a Bishop Quentin Huang. He was a product of modern Chinese, Westernized education. He had small weekly services for the professional Chinese community, most of whom were scientifically oriented; students, engineers, professors, physicians and scientists. His perennial theme was that religion, specifically Christianity, did not conflict with science. To me, harping on this theme was a wasted effort. Having been trained in philosophy, specifically in the philosophy of science, I knew that scientific propositions do not contain all of knowledge. Propositions concerned with religion are by definition not scientific, because according to Popper, they cannot be negated. Religions answer basic questions of man, about the nature and meaning of life, of reality, and of the universe. They are questions that cannot be answered completely by the methods of science. But that does not mean they are a lesser component of human knowledge.

It is true that some of these questions that used to belong to religion have been answered by science but they have been replaced by others. Europeans used to think that the sun revolved around an immovable earth. This law was thought to be enshrined in the Bible and was inviolable. As late as 1633, Galileo was declared a heretic by the church because he taught that the earth revolved the sun. In 2001, the Catholic Church apologized for having persecuted Galileo. The present scientific theory of the beginning of the universe is the big bang theory. But these answers are still only partial ones. The theory tells us what happened billions and billions of years ago. One can still question what happened before the so-called beginning.

Other questions have been answered by science beyond the ken of previous imagination. The nature of life is one such. We now know more about the physical and chemical basis of life than we ever imagined. But in a sense, these answers raise more questions than they answer. They have not reduced the wonder and marvel of life one *whit*. To me the movement of inanimate molecules in carrying out the basic processes of life is so remarkable that I often shudder in wonderment. In a way the creation and purpose of life remains as inscrutable as before we knew anything about its science. Therefore the nature of religious questions changes as science advances, but the number and depth of such questions are not reduced. To me questions of science and of religion are of different orders, although there is mixing and exchange on the borders between the two fields. Science cannot answer the basic religious question of the purpose and meaning of life. Throughout the ages, this remains the "mysterium tremendum"!

There is a naive streak in modern Western education in China that posits the simple antagonism between science and religion. This streak is greatly re-enforced by the materialism of Marxism, which has no use for religion and considers it the "opium" of the people. Under this formulation, religions are reduced to superstitions. Marxism was a powerful intellectual force in China even before the establishment of communism in 1949. While it is now discredited, and religions remain alive and well, it has left its mark on the misunderstanding of religions among Chinese intellectuals.

There is also another streak in the Chinese intellectual tradition that denigrates religions. To the Chinese mind, totally devoid of the entrenched, rich traditions of Christianity in the West, religions and superstitions, are not clearly distinguished. The traditional intellectual Chinese mind was dominated by Confucianism, which is basically agnostic with regard to religion. Confucius was the supreme humanist. His views regarding religion could be summarized in his famous aphorisms: "Respect the gods and devils, but distance yourselves from them"; "If you do not know Life, how can you know Death". It seems that Confucius, as the leading Chinese intellectual over the centuries, in his preoccupation with the proper ordering of man's human relations in wordly existence, squeezed out all interest in metaphysical questions. They were posed by some of his contemporary thinkers like Chuantse, Laotse and Motse. But their views were outside the mainstream.

The total absence of teaching about religions in China is an anomaly. In particular the total negligence of Buddhism in modern Chinese education is a great defect. Buddhism after all was and is an important fact in Chinese culture and history. I don't remember hearing a single lecture on Buddhism in all my classes in middle school and in college in China. Buddhism has been a profound and abiding influence in China since before the birth of Christ, down through each of the major dynasties to the present day. It has influenced the Chinese language, art, thinking and literature. It is astounding to me that none of this was thought worthy to be taught! I ascribe it as a blind spot in Chinese education.

It is only after my retirement that I began to be interested in Buddhism. Before going to Taiwan in 1997, I read a small volume by the Dalai Lama, the title of which is more striking than its contents, *How to Live and How to Die*[121]. Unlike Confucius, his point was that it is as important to know how to die as it is to know how to live. This impressed me, because I was disturbed by my father's fear of death. How to deal with death is certainly a major problem of all great religions.

O Death, where is thy Sting?

"Cho-Cho died at 5:10 a.m." These were the stark words which greeted me as I read a telephone message. It was 8:30 a.m. on March 28 (my birthday), 1981. The place was Wilson Lodge, Oglebay Park, Wheeling, West Virginia. I was attending a weekend meeting of the Executive Committee of the Department of Pathology. I had just come back from a refreshing walk through the beautiful grounds of this scenic resort near Pittsburgh. The weather was glorious. It was the first real spring day of the year.

Could those scribbled words really be true? I called home. Bettie and Carol confirmed the worst. On March 26, I had fed Cho-Cho some steak we were having for dinner. We don't eat steak much any more, and Cho-Cho devoured the morsels I gave her voraciously. We knew she had a delicate digestive system. After too much chicken or meat, she had a tendency to throw up. I tried to give her as little as possible. She was so eager I probably gave her too much. During the night and next morning, Cho-Cho looked unhappy. In retrospect, she was mortally ill and I did not realize it. She had vomited little bits on the rug, but I did not think too much of it because it had occurred so often before. She was losing precious body fluid. During the afternoon of March 27, I saw her briefly before leaving for Oglebay. I was surprised she had not recovered. I should have taken her to the veterinarian to have her fluids replaced intravenously. She was the size of a human infant who cannot tolerate fluid loss. But I did not realize this either. Bettie came home and did finally take her to the animal hospital that night. But it was too late. I am glad Bettie tried. She was always more conscientious about taking Cho-Cho to the vet.

That tiny island of life, Cho-Cho, is dead. I tried to comprehend this the rest of the day. I sat at the meeting and listened to the torrent of words. But my mind wandered. Thoughts and images of Cho-Cho kept appearing and reappearing.

I always knew that Cho-Cho would one day die. This tiny bundle of joy was too marvelous to last very long. Carol said she was like a wound-up toy. What a magical toy: A toy with feelings. Every morning she would greet us upon awakening, after winding up by stretching. She delighted in her daily routine. She was eager to go out in the morning. She was eager to come back. She had a sixth sense about our intentions. When we went off to work, she would already be lying languidly on the stairs, observant but uninterested, getting ready for her day long siesta while we were at work. But when she was included in an outing on holidays and weekends, she would instinctively know. Jumping up and down, she was eager to go.

I was the one who took Cho-Cho out everyday, at least twice a day when I was home. I took her out in rain or shine, snow or wind. By all odds, I should have tried to get others to share in a boring chore. But I did not. In fact, I felt put out if some one else in the family took her out in my stead. Some times I wondered why this was so. I think it was because I was fascinated by her being alive. Her physiologic functions assured me that this living machine was functioning properly. I enjoyed fully each day of her short life.

I will miss most of all her greeting me everyday when I came home from work. After I parked the car in the garage, and as I walked up the driveway, here was this tiny wisp charging in my direction to greet me. Carol had thoughtfully opened the front door to let her out for this daily ritual. There was some enigmatic logic by which she decided how she would greet people. When we returned from Australia after an absence of a whole year, I anticipated her greeting us with delirious enthusiasm. Instead, she barely acknowledged our return. Perhaps she was chiding us for being away so long. On the other hand, whenever the children or Uncle Albert returned after being away for weeks or even months, her greetings bordered on the ecstatic.

Each member of the family will miss Cho-Cho in his own way. Bettie and Didi (John) will remember her sleeping on their beds, something Carol and I would not tolerate. Whenever Carol, I and Cho-Cho were relaxing in the living room, Cho-Cho always chose Carol to snuggle up to instead of me. She would lie for hours beside Carol in her chair. Carol will miss this.

Being a pedigreed Maltese, Cho-Cho might be expected to be nervous and high strung. She was actually easy-going and well mannered. She had so few needs; she did not even shed any hair in the summer. We had a five inch high board to keep her out of the living room when we were away. She could have easily jumped over it. Typically, she had the good taste never to do that. At the tender age of one or two months when she first arrived, we hardly had any trouble toilet training her. During the first year, she chewed on some shoe laces, but soon gave that up. She did retain the curious habit of overturning our waste paper baskets. This little mischief was small price to pay for her general good behavior. We simply kept our baskets emptied.

Friends and relatives alike doted on Cho-Cho. During her entire life, we never had to board her in a kennel. The family of Colleen Kennedy (family domestic) loved her as we did, and took her in when we had to go out of town. They too will miss her. She was the center of attention during the Ho family reunion at our house last Christmas. Allocating her affections equally, she would sit close to each member of the family in turn.

Cho-Cho was tiny; a mere five pounds of silky white hair. One time, she was mistaken for a handkerchief dropped by an elderly lady friend. She was delicate, as her rapid demise showed. But she had a frisky robust disposition. Her bark was strong and manly despite her size. Oddly, I never thought of her as a female. I was constantly referring to her as "him". Actually she was more like an "it". Sex has little meaning outside of one's own species. Cho-Cho spent all her life with us humans. I used to wonder if she missed her own kind. But I am convinced dogs are closer to man than to dogs.

The talk at my meeting droned on. I tried to remember the year Cho-Cho was born in. We got her as a present for Bettie when she made honor roll in high school. She was born on February 29, clearly in a leap year. Leap years are even years. So she must be born either in 1972 or 1974. Not getting any further, I asked a colleague next to me when the last leap year was. He said 1980, and that leap years are always presidential election years. I figured Cho-Cho was born in 1972, nine years ago. I felt relieved that it was not 1974. Otherwise she would have died too young. As it is, if one year of a dog's life is equivalent to seven human years, she died at the age of 63. That is not old, but ripe. It is not a bad age to die. And die we must all.

Death is part of life. It is unpleasant because it is the great Unknown. Part of life is to come to terms with death, even if we'll never understand it or accept it fully. Cho-Cho is dead. That is such an unalterable fact. She leaves a gap in our lives. It cannot be filled. We will always remember her and long for her, even though the sands of time will blunt the pain of our loss.

I continued to doodle on a piece of paper as the above thoughts raced through my mind during the mercifully long meeting. Towards evening, I doodled the following inscription for her hypothetical gravestone:

HERE LIES CHO-CHO HO

1972–1981

MAY HER SOUL REST IN PEACE

All her life she gave and received Love

Then I thought: To give and receive love, would that not even be fulfillment for a human life? Should any living creature wish for more? Cho-Cho was one of God's very own for being so blessed. I envy the perfection of her tiny life.

After Cho-Cho died, Bettie Pei-wen planted in her memory a small tree in our back yard. For the first few years, I was concerned that it would not survive. It was struggling for sunlight in the shade of its neighbors. But today, more than twenty years later, this tiny tree has grown up to over thirty feet. It is flourishing. It is overwhelming in its exuberance. Looking at it, I remember Cho-Cho, but it no longer has any resemblance to her. It is another life, totally different and separate. This tree almost makes me forget Cho-Cho. Is this the meaning of everlasting life?

After going to Taiwan, I began to listen to a Buddhist monk called Ching Kun Fa Tse on T. V. He gives a sermon, some of which are taped, every single day at 6:30 a.m. I listened to him almost five years.

One of his aphorisms is that "Buddhism is not a religion, Buddhism is not a philosophy, Buddhism is a system of education". Like all aphorisms, it has elements of truths. Buddhism is probably all of these things, but its emphasis on "how to" underlies its being a system of education. The preoccupation of Buddhism is "how to" achieve enlightenment, or wisdom. It teaches that our true being or enlightened self is clouded by our desires, avarice, worries, selfishness, preoccupations, intemperance, and arrogance. How to divest ourselves of these encumbrances is the goal of proper living. It is interesting to compare Buddhist charity or love with the Christian counterpart. Buddhism teaches charity and philanthropy as a method to become disencumbered. It is an expression of pity. It is a means to an end. In Christianity love is the end itself, since it comes from God.

A remarkable aspect of Buddhism is its metaphysics, or understanding of reality beyond the physical world. All physical phenomena including mental processes are ephemeral and transitory. However there is an indescribable quality about life, which is eternal and may take form in plants, animals or in man. This is the basis of the reincarnation theory. A human's life can be reincarnated in an animal after death. Reincarnations are endless cycles. Only enlightenment, or true understanding, can stop this process. I find this part of Buddhism difficult to accept, just like the Apostle's Creed of Christianity. But it does have a positive aspect that is not usually emphasized. Reincarnation is the basis for Buddhism's "reverence for life", a doctrine which Albert Schweitzer, one of the great Christians of the twentieth century, thought he discovered late in his rich life. It was the basis of his many beliefs. Actually this doctrine had been inherent in Buddhism for more than two thousand years before he "discovered" it. I find this coincidence remarkable. This shows that great religious truths are constantly being rediscovered, since they are expressions of sublime elements in human nature.

Buddhism does not postulate a creator, a prime mover of the world, or God or gods that can act beyond physical laws. Questions such as "creation", that have haunted western philosophy and religions for centuries are no concern to Buddhism. Of all religions, Buddhism is the only one that teaches the universe is governed by strict laws of cause and effect. A supernatural god or creator is unnecessary. This is the reason why Ching Kun says Buddhism is not a religion. In this sense it is remarkably modern. The path to wisdom or enlightenment is preceded by a constant building up of the causal foundations of the good life, which consists of successive detachment from our specific encumbrances.

Conclusion

The understanding of religions has been an important part of my education. Religious questions are legitimate, important concerns of man. But they cannot be solved by the usual methods of science. The other method is acceptance of its teachings by a "leap of faith". Those for whom this is possible are in a way fortunate. For they become assured in the answers and convictions. And they may have peace.

I have taken the route of exploring the teachings of great religions rather than trying to believe them. So far I have limited myself to Christianity and Buddhism. While I cannot completely accept all the beliefs of either religion, I feel each provides important answers and insights. I realize that snatches of insight do not constitute a systematic code of beliefs, so in a sense I have no religion. Still these universal insights have been useful to me at the personal level.

I do believe that there is a spiritual life. Meditation is one way to enhance it. I have made some progress in my spiritual life. But there is much to learn. It is a type of knowledge that can improve.

The idea of an almighty God has been one of the foundations of belief in Western civilizations for almost two thousand years. It has been so ingrained that it has not occurred to many in the West to even question it. It solves many questions of religion. The big problem about a God is its postulation in the first place. One way to look at God is to consider it as the "inner light", as George Fox taught the Friends. This concept is akin to Buddhist enlightenment.

Buddhism, a great religion of the East, shows us that monotheism is not essential to be a great religion. It is no accident that the most important religions were developed by great human beings. Buddhism was developed by Shakyamuni, whose long life of prolific teaching was in striking contrast to Jesus' mere thirty years. Buddhism teaches many ways to reach enlightenment. Again in striking contrast to dogmatism and fanaticism of monotheism, there are 840 such ways. Buddhism is a bit like an encyclopedia of "how to". The common theme is that the basis of non-enlightenment is preoccupation with selfish desires, wealth and personal well being. This is Buddhism's lasting insight. Still when all is said and done, enlightenment is very much a mystical state, a religious state, much desired but rarely achieved.

Where does all this leave me, the explorer? I certainly have not reached nirvana, nor is it likely that I will in my lifetime. I do not find peace in the idea of a God who is an omnipotent creator. But I do like the idea that God is Love.

To understand all aspects of love and to practice it beyond one's own is worth a life's effort. Very likely, this is about all that can be expected of some modern men. The great questions of religion remain unanswered. That too may be part of being human. And I too may fear death when my time comes. I do not consider that unnatural.

chapter

| 20 |

AMERICA

The first thirty years of my life were peripatetic and filled with move-ment; living and going to schools in three different continents: six years spent in schools and colleges in widely separated areas of war-torn China and finally completing my formal education in the United States. The sec-ond forty-five years of my life have been remarkably sedentary. Except for a few years after retirement, my entire academic medical career has been at the University of Pittsburgh.

I am thankful for the richness and happiness of my life, most of which has been in the United States. Despite the complexity of my background and the diverse components of my personality, the foundation for my career was laid at Harvard University, an institution to which I feel more and more gratefulness as I age. In my contribution to the 25th anniversary report of our class of 1949, I wrote "I echo Henry Kissinger, another Harvard graduate, when he, born a German Jew, was sworn in as Secretary of State, said, 'It could have only happened in the United States' ".

Not for naught is America held up as the golden land of opportunity. My classmate, Hu Xiao-chi, wrote me from Beijing after reading the Chinese version of this book that she was happy I emigrated from China when I did, for those like her who remained in China and underwent the trials and travails of the Cultural Revolution had no opportunity to accomplish as I did. Their opportunities were irretrievably lost.

America is a diverse country. One is free to live any way one wishes. Still, this maxim is only partly true. Ordinary Americans are remarkably similar culturally. They have the same likes and dislikes. Fads and fashions are shared in common and are widespread and pervasive. In the 1950s, when I was attending Stanford University, I became shockingly aware of the undercurrent of prejudice against the Chinese. Such prejudice was most acutely discernible in the taboo against interracial dating. One could feel it without being informed. I found the uniformity of the Californian college students offensive. When we attended a Stanford football game, we had to sit in the Stanford section in a spotless white shirt. Social groups at the university or the medical school were stereotyped and did not necessarily welcome the Chinese. I found the native born Chinese unbelievably submissive and withdrawn. They were like downtrodden people! I recall being surprised at the frank materialistic aspirations of a fellow graduate student with whom I was conversing. I found no trace of idealism.

I realize nowadays that racial prejudice against the Chinese has all but disappeared in California as well as the rest of the country. In fact, there is "reverse prejudice" in the sense that in elite academic institutions like Berkeley large numbers of Chinese students inadvertently enhance the standards of certain classes, and exclude other students. This remarkable change I credit to crusaders like Martin Luther King. Their successful fight for civil rights for African Americans has benefited minorities like the Jews, Chinese and Japanese more than the African American. I feel grateful to them.

Still, if I am asked, "Do you really feel at home in the United States?" My answer would be a conditional yes. I am at home here because I have been here for over fifty years and I could not have hoped for a better home. Yet I know that if I were to engage in a casual conversation with a car repair man, as happened just a year ago, I can sense the eagerness of his inevitable question, "How long have you been living here?" It should be of satisfaction to me to be able to answer, "I have been living here longer than you have been alive." But irrespective of how long I have lived here, to him, and indeed to myself, by dint of my appearance and origin, I will always in a sense be a foreigner.

As I have explained elsewhere in this book, I was late in my life able to satisfy my allegiance to my non-American origin, by working for and with the Chinese people for a few years. I consider myself fortunate by training and ability to be able to do this. I feel in many ways like a citizen of the world. I am proud that being an American allows me to be "diverse" in this way. But what about those who are dissatisfied by being considered diverse? What if they will be considered foreign even if they feel themselves to be 100% American? This is the ultimate racial question in this country, the answer to which is not yet clear. America used to be a country where

"WASP's" (white, Anglo-Saxon, Protestant, now a jaded term) were advantaged. This is because in the last hundred years, discrimination against the Irish, Italians, Germans, Catholics and Jews has disappeared one by one. Hopefully that means that equality toward all other ethnic, racial, religious and cultural groups will be forthcoming. But this cannot be taken for granted.

The American Aspiration

Two remarkable documents underscore America's political aspirations; the Declaration of Independence (1776), and the U.S. Constitution (1787). They are the bases of American government and they are what Americans mean when they advocate racial equality and human rights among nations, and "democracy" throughout the world. To Americans, a democracy is not so much a rule of the people as a government of representatives elected by the people. It is no accident that such representative "democratic" governments, similar to what is described by the U.S. Constitution, were created under American influence and pressure in the Philippines and in South Vietnam, after World War II. Whether such "democracies" really represented the people or special interests is a moot point. Similarly, over the last two hundred years, America has been the model for the governments of Latin America. Not many of them have been a smashing success. What has been overlooked is that these countries do not have the cultural and social perquisites that make such democratic institutions workable. Therefore a serious student of politics will have to consider America's political aspirations for the world somewhat superficial. At times it seems to emphasize form more than substance.

Still, America's ideas were powerful for the world during World War I, World War II and the Cold War. American ideals were perfect antidotes against imperialism, colonialism, monarchies, and the Nazi and Russian dictatorships. Is it then a wonder that America, now unwittingly the world's sole superpower, should consider its ideals and aspirations also supreme for the world? If we are in the era of "Pax Americana", America is in an unusually good position to be the world leader. Americans are believed by nature to be non-aggressive since they already possess most of the wealth of the world. Peoples from all over the world have immigrated to the United States, and bear witness and spread the gospel of its munificence. Alone among the Western nations, the United States has stood for self determination of peoples, and against the colonizing imperialistic powers.

However the world has changed now that the United States is the sole superpower. During the Cold War, the United States had to deal with another superpower, and its policies had to be in part dictated by "Realpolitik", a specialty of our previous Secretary of State, Henry Kissinger. Now that there is no power to counterbalance that of the United States, America's outlook and strategy must accordingly change.

Let me just make two points. For almost one hundred years now, the United States has fostered first the League of Nations, and then the United Nations as institutions to preserve world order. Now that there is no one to oppose us, the U.S. has the responsibility to see that the UN works. It can no longer afford to ignore it because without the UN, the U.S. would have to create something like it in order to police the world. Members of the Congress should realize this and support the UN rather than attacking it. If they don't like parts of it, they can try to change it.

By the same token, the U.S. should accept the principle that no international police actions should take place outside the UN. That was its original purpose, and it is the only institution we have to carry out this purpose. It would be contrary to the intentions of the founders of the UN, the most important of whom were Americans, if the U.S. invaded another country without the consent of the UN because it was labeled by the President of the United States as "evil".

Second, despite America's being the sole superpower, the principle of "sovereignty" is still paramount among nations. It is apparent that the United States itself is extremely jealous of its sovereignty. Some members of Congress seem to object to the UN because it violates the sovereignty of the United States. Surely in an organization of independent states, we have the obligation to respect the sovereignty of other states. We cannot afford to violate the sovereignty of another nation without "due process", which in the world of today, is action by the UN. Unless the United States leads by following proper rules of international behavior, there would be international chaos.

Historically, the United States has never attacked another state without provocation outside of the Western hemisphere, where the Monroe Doctrine applies. President George W. Bush attacked Iraq preemptively, without clear evidence that it threatened the security of the United States, and without UN consent. This double error damaged America's position as a responsible world leader. The sovereignty of a Muslim state on another continent was violated. Of all nations of the world, the Muslim nations have more just grievances against the United States than others, because of our slanted support for Israel. Despite and indeed because of 9/11, how the U.S. deals with a Muslim state requires an extra ounce of discretion and statesmanship. Otherwise there is a true danger of a clash of civilizations, the Muslim world against the United States. One of the most naïve current ideas is that Americans can walk into Baghdad and change the whole Muslim world by making Iraq "democratic". Such a mission smacks of imperialism. It is like the "white man's burden" of 19th century European colonialism.

I feel that President Bush has violated the basic ideals of American diplomacy of the last hundred years. Hopefully the built-in mechanisms of democratic correction will offset the harm done. But at this time in early 2005, it is difficult to see how we are going to get out of Iraq.

America is a relatively young and generous nation, uniquely blessed by nature and endowed with powerful ideals. It has been my good fortune to have become an American. I am proud to be one. By the same token, as a citizen of this remarkable democracy, I consider the responsibility of being an American seriously. And that includes participating in formulating its future, which is replete with challenges, responsibility and hope.

APPENDIX

APPENDIX

REFERENCES AND NOTES

Author's Preface

1. Ho M. *My Education, My Path in Medicine*. Taipei, Taiwan: The Journalist Publishing Company; 2002. 285 pages (in Chinese).

Chapter 1

2. Ho FS. *Forty Years as a Diplomat*. Hong Kong: Chinese University Press; 1990. 713 pages (in Chinese).

Chapter 4

3. Yad Vashem, The Holocaust Martyrs' and Heroes' Remembrance Authority [Internet]. Jerusalem: Yad Vashem; 2004; Available from: *http://www.yadvashem.org/*

Chapter 5

4. Marx K, Engels F. The Communist Manifesto. 1948.
5. Popper K. *The Open Society and its Enemies*. 4th ed. Princeton, NJ: Princeton University Press; 1963.
6. Popper K. *Logik der Forschung, zur Erkenntnistheorie der Naturwissenschaft*. Vienna: Julius Springer; 1935.
7. Grünbaum A. *Foundations of Psychoanalysis: A Philosophic Critique*. University of California Press; 1984.
8. Mannheim K. *Diagnosis of our Time*. Oxford University Press; 1944.
9. Mannheim K. *Ideology and Utopia, An Introduction to the Sociology of Knowledge*. Translated by Louis Wirth and Edward Shils. New York: Harcourt, Brace and Co.; 1936.

Chapter 6

10. Tsu YY. *Friend of Fisherman*. Fort Washington, PA: Trinity Press; 1968.
11. Huie Kin. *Reminiscences*. Beijing, China: San Yu Press; 1932.

Chapter 7

12. Snapper I. *Chinese Lessons to Western Medicine; a Contribution to Geographical Medicine from the Clinics of Peking Union Medical College*. New York: Grune and Stratton; 1965.
13. Ferguson ME.*China Medical Board and Peking Union Medical College*. China Medical Board of New York, Inc.; 1970.
14. Bowers JZ. *Western Medicine in a Chinese Palace*. Peking Union Medical College, 1917–1951. Philadelphia: Josiah Mary, Jr. Foundation; 1972.
15. Flexner A. *An autobiography*. Simon and Schuster; 1960.
16. Flexner A. Medical Education in the U.S. and Canada. A Report of the Carnegie Foundation for the Advancement of Teaching. A study of 155 medical schools. 346 pages. 1910.
17. Kass EH. Asymptomatic infection of the urinary tract. *Tr Assoc Am Physi*. 1956;69:56–64.
18. Ho M, Kass EH. Protective effect of components of normal blood against the lethal action of endotoxin. *J Lab Clin Med*. 1958;51:297–311.
19. Enders JF, Weller TH, Robbins FC. Cultivation of the Lansing strain of poliomyelitis virus in cultures of various human embryonic tissues. *Science*. 1949;109:85–87.
20. Enders JF, Peebles TC. Propagation in tissue culture of cytopathogenic agents from patients with measles. *Proc Soc Exp Biol Med*. 1954;86:277–286.
21. Price L. Dialogues of Alfred North Whitehead, as recorded by New American Library; 1956.
22. Robbins, FC. From philology to the laboratory. *Harvard Medical Alumni Bulletin*. Winter 1985; 16–18.
23. Isaacs A, Lindenmann J. Virus interference I. the interferon. *Proc Royal Soc B*. 1957;147:258–267.
24. Ho M, Enders JF. An inhibitor of viral activity appearing in infected cell cultures. *Proc Natl Acad Sci USA*. 1959;45:385–389.
25. Ho M, Enders JF. Further studies on an inhibitor of viral activity appearing in infected cell cultures and its role in chronic viral infections. *Virology*. 1959;9:446–477.
26. Ho M. An early interferon: Viral inhibitory factor. *J Interferon Res*. 1987; 7:455–458.

Chapter 8

27. Krim M, Came PE, Carter WA. *Interferons and the Applications.* Berlin: Springer; 1984: pp. 1–15.

28. Strander H. Anti-tumor effects of interferon and its possible use as an antineoplastic agent in man. *Texas Reports on Biology and Medicine.* 1977;35:429–435.

29. Holland JJ, McLaren LC, Syverton JT. The mammalian cell-virus relationship. IV. Infection of naturally insusceptible cells with enterovirus RNA. *J Exp Med.* 1959;110:65–80.

30. Ho M. Inhibition of the infectivity of poliovirus ribonucleic acid by an interferon. *Proc Soc Exp Biol Med.* 1961;107:639–644.

31. Ho M. Kinetic considerations of the inhibitory action of an interferon produced in chick cultures infected with Sindbis virus. *Virology.* 1962;17:262–275.

32. Wheelock EF, Dingle JH. Observations on repeated administration of viruses to a patient with acute leukemia. *New Eng J Med.* 1964;271:645–651.

33. Gledhill AW. Sparing effect of serum from mice treated with endotoxin upon certain murine virus diseases. *Nature.* 1959;183:185–186.

34. Ho M, Kass EH. Protective effect of components of normal blood against the lethal action of endotoxin. *J Lab Clin Med.* 1958;51:297–311.

35. Ho M. Interferon-like viral inhibitor in rabbits after intravenous administration of endotoxin. *Science.* 1962;146:1471–1472.

36. Stinebring WR, Youngner JS. Patterns of interferon appearance in mice injected with bacteria and bacterial endotoxin. *Nature.* 1964;204:712–715.

37. Field AK, Tytell AA, Lampson GP, Hilleman MR. Inducers of interferon and host resistance. II. Multistranded synthetic polynucleotide complexes. *Proc Natl Acad Sci (US).* 1967;58:1004–1010.

38. Ho M, Kono Y, Breinig MK. Tolerance to the induction of interferons by endotoxin or virus. *Proc Soc Exp Biol Med.* 1965;119:1227–1232.

39. Billiau A. The refractory state after induction of interferon with double-stranded RNA. *J Gen Virol.* 1970;7:225–232.

40. Breinig MK, Armstrong JA, Ho M. Rapid onset of hyporesponsiveness to interferon induction on reexposure to polynucleotides. *J Gen Virol.* 1975;26:149–158.

41. Cantell K. The Story of Interferon, the Ups and Downs in the Life of a Scientist. Singapore: *World Scientific*; 1998.

42. Youngner JS, Stinebring WR, Taube SG. Influences of inhibitors of protein synthesis on interferon formation in mice. *Virology.* 1965;27:541–564.

43. Ho M, Kono Y. Effect of actinomycin D on virus and endotoxin-induced interferon-like inhibitors in rabbits. *Proc Natl Acad Sci USA.* 1965;53:220–224.

44. Ho M, Ke YH. The mechanisms of stimulation of interferon production by a complexed polyribonucleotide. *Virology.* 1970;40:693–702.

45. Ho M, Ke YH, Armstrong JA. Mechanisms of interferon induction by endotoxin. *J Infect Dis.* 1973;128:220–227.

46. Tan YH, Armstrong JA, Ke YH, Ho M. Regulation of cellular interferon production: Enhancement by antimetabolites. *Proc Natl Acad Sci USA.* 1970;67:464–471.

47. Taniguchi T, Buarente L, Roberts TM, et al. Expression of the human fibroblast interferon gene in *E. coli. Proc Natl Acad Sci USA.* 1979;77:5230–5233.

48. Ho M, Tan YH, Armstrong JA. Accentuation of production of human interferon by metabolic inhibitors. *Proc Soc Exp Biol Med.* 1972;139:259–262.

49. U.S. Patent No. 3773924.

50. Soloviev VD. Some results and prospects in the study of endogenous and exogenous interferon. In G. Rita, *The Interferons.* New York: Academic Press; 1968:233–243.

51. Merigan TC, Reed SE, Hall TS, Tyrrell DA. Inhibition of respiratory virus infection by locally applied interferon. *Lancet 1.* 1973;803:563–567.

52. Pazin GJ, Ho M, Janetta PJ. Reactivation of herpes simplex virus after decompression of the trigeminal nerve root. *J Infect Dis.* 1978;138:405–409.

53. Pazin GJ, Armstrong JA, Lam MT, Tarr GC, Janetta PJ, Ho M. Prevention of reactivated herpes simplex infection by human leukocyte interferon after operation on the trigeminal root. *N Engl J Med.* 1979;301:225–230.

54. Ho M, Pazin GJ, Armstrong JA, Haverkos HS, Dummer JS, Janetta PJ. The paradoxical effects of interferon on reactivation of oral herpes simplex infection after microvascular decompression for trigeminal neuralgia. *J Infect Dis.* 1984;150:867–872.

55. Pazin GJ, Harger JH, Armstrong JA, Breinig MK, Caplan RJ, Cantell K, Ho M. Leukocyte interferon for treating first episodes of genital herpes in women. *J Infect Dis.* 1987;156:891–898.

56. Ho M. Interferon as an agent against herpes simplex virus. *J Invest Dermatol.* 1990;95:158S–160S.

57. Ho M, Pazin GJ, White LT, et al. Intralesional treatment of warts with interferon-alpha and its long term effect on NK cell activity. In E. DeMaeyer, G. Galasso and H. Schellenkens, eds. *The Biology and the Interferon System.* Elsevier/North Holland Biomedical Press; 1981.

58. Pazin GJ, Ho M, Haverkos HW, Armstrong JA, Breinig MK, Wechsler HL, Arvin A, Mergian TC, Cantell K. Effects of interferon-alpha on human warts. *J Interferon Res.* 1982;2:235–243.

59. Gui XE, Ho M, Cohen MS, et al. Hemorrhagic fever with renal syndrome: Treatment with recombinant alpha interferon. *J Inf Dis.* 1987;155:1047–1051.

60. Quesada JR, Reuben J, Manning JT, Hersch EM, Gutterman JV. Alpha interferon for induction of remission in hairy-cell leukemia. *New Engl J Med.* 1984;310:15–18.

61. Krown SE. The role of interferon in the therapy of epidemic Kaposi's sarcoma. *Semin Oncol.* 1987;14(suppl 3):27–33.

Chapter 9

62. Starzl TE. *The Puzzle People, Memoirs of a Transplant Surgeon.* Pittsburgh and London: University of Pittsburgh Press; 1992. (364 pages.)

63. Moore FD. *A Miracle and a Privilege, Recounting a Half Century of Surgical Advance.* Washington, DC: Joseph Henry Press; 1995. (450 pages.)

64. Ho M, Jaffe R, Miller G, Breinig MK, Dummer JS, Makowka L, Atchison RW, Karrer F, Nalesnik MA, Starzl TE. The frequency of EBV infection and associated lymphoproliferative syndrome after transplantation and its manifestations in children. *Transplantation.* 1988;45:719–727.

65. Hanto DW, Sakamoto K, Purtilo DT, Simmons RL, Najarian JS. The Epstein-Barr virus in the pathogenesis of posttransplant lymphoproliferative disorders. Clinical, pathologic, and virologic correlation. *Surgery.* 1981;90:204–213.

66. Starzl TE, Nalesnik MA, Porter KA, Ho M, et al. Reversibility of lymphomas and lymphoproliferative lesions developing under cyclosporine-steroid therapy. *Lancet.* 1984;1:583–587.

67. Ho M, Miller G, Atchison RW, Breinig MK, Dummer JS, Andiman W, Starzl TE, Eastman R, Griffith BP, Hardesty RL, Bahnson HT, Hakala TR, Rosenthal JT. Epstein-Barr virus infections and DNA hybridization studies in post-transplantation lymphoma and lymphoproliferative lesions: Role of primary infection. *J Inf Dis.* 1985;152:876–886.

68. Cen H, Breinig MC, Atchison RW, Ho M, McKnight JLC. Epstein-Barr virus transmission via the donor organs in solid organ transplantation: PCR and RFLP Analysis of IR2, IR3 and IR4. *J Virol.* 1991;65:976–980.

69. Weller TH, Macauley JC, Craig JM, et al. Isolation of intranuclear inclusion producing agent from infants with illnesses resembling cytomegalic inclusion disease. *Proc Soc Exp Biol Med.* 1957;94:4–12.

70. Ho M. *Cytomegalovirus: Biology and Infections,* Second Edition. New York: Plenum Medical; 1991. (440 pages.)

71. Healey-White ET, Craighead JE. Generalized cytomegalic inclusion disease after renal homotransplantations — report of a case with isolation of virus. *N Engl J Med.* 1965;272:473–475.

72. Ho M, Suwansirikul S, Dowling JN, Youngblood LA, Armstrong JA. The transplanted kidney as a source of cytomegalovirus infection. *N Engl J Med.* 1975;293:1109–1112.

73. Suwansirikul S, Rae N, Dowling JN, Ho M. Primary and secondary cytomegalovirus infection. *Arch Intern Med.* 1977;137:1026–1029.

74. Meyers JD, Flournoy N, Thomas ED. Risk factors for cytomegalovirus infection after human marrow transplant. *J Infect Dis.* 1986;153:478–488.

75. Dummer JS, Armstrong J, Somers J, Kusne S, Carpenter BJ, Rosenthal JT, Ho M. Transmission of infection with herpes simplex virus by renal transplantation. *J Infect Dis.* 1987;155:202–206.

76. Dowling JN, Saslow AR, Armstrong JA, Ho M. Cytomegalovirus infection in patients receiving immunosuppressive therapy for rheumatologic disorders. *J Infect Dis.* 1976;133:399–408.
77. Ho M. Interferon. In: 1964, 1970 eds. McGraw-Hill Year Book of Science and Technology. Page 255. 1970.
78. Dowling JN, WU BC, Armstrong JA, Ho M. Enhancement of murine cytomegalovirus infection during graft versus host response. *J Infect Dis.* 1977;135:990–994.

Chapter 10

79. Finland M. *The Harvard Medical Unit at the Boston City Hospital.* (History of the Thorndike Memorial Laboratory and the Harvard Medical Services from their Founding until 1974). In three volumes. Volume 1, 903 pages. University of Virginia Press, 1982.
80. Ho M, Ashman RB. Development in vitro of cytotoxic lymphocyte against murine cytomegalovirus. *Aust. J Exp Biol Med Sci.* 1979;57:425–428.
81. Ho M. Role of specific cytotoxic lymphocytes in cellular immunity against murine cytomegalovirus. *Infect Immun.* 1980;27:767–776.

Chapter 11

82. Blockenstein Z. Graduate School of Public Health, University of Pittsburgh, 1948–1974. 226 pages. A. W. Mellon Educational and Charitable Trust, Pittsburgh, PA. See page 156.
83. WHO. Declaration of global eradication of small pox. *Weekly Epidemiol Rec.* 1980;55:145–152.
84. Lederberg J, Shope RE, Oaks SC Jr, eds. *Emerging Infections: Microbial Threats to Health in the United States.* Washington, DC: National Academy Press; 1992.

Chapter 14

85. Schmidt NJ, Lennette EH, Ho HH. An apparently new enterovirus isolated from patients with disease of the central nervous system. *J Infect Dis.* 1974;129:304–309.
86. Shindarov LM, Chumakov MP, Voroshilova MK, et al. Epidemiological, clinical, and pathomorphological characteristics of epidemic poliomyelitis-like disease caused by enterovirus 71. *J Hyg Epidemiol Microbiol Immunol.* 1979;23:284–295.
87. Nagy G, Takatsy S, Kukan E, Mihaly I, Domok I. Virological diagnosis of enterovirus type 71 infections: experiences gained during an epidemic of acute CNS diseases in Hungary in 1978. *Arch Virol.* 1982;71:217–227.

88. Landry ML, Fonseca SNS, Cohen S, Bogue CY. Fatal enterovirus type 71 infection: rapid detection and diagnostic pitfalls. *Pediatr Infect Dis J.* 1995;14:1095–1100.

89. Baker AB. Polomyelitis. 16. A study of pulmonary edema. *Neurology.* 1957;7:743–751.

90. Hammon WM, Rudnick A, Sather GE. Viruses associated with hemorrhagic fevers of the Phillipines and Thailand. *Science.* 1960;131:1102–1103.

91. Chang LY, Huang YC, Lin TY. Fulminant neurogenic pulmonary oedema with hand, foot, and mouth disease. *Lancet.* 1998;352:367–368.

92. Ho M, Chen ER, Hsu KH, Twu SJ, Chen KT, Tsai SF, Wang JR, Shih SR. The Enterovirus Type 71 epidemic of Taiwan, 1998. *N Engl J Med.* 1999;341:929–935.

93. Salk JE, et al. Formaldehyde treatment and safety testing of experimental poliomyelitis vaccines. *Am J Pub Health.* 1954;44:563–570.

94. Francis T, Jr, et al. Evaluation of the 1954 field trial of poliomyelitis vaccine (final report), Poliomyelitis Vaccine Evaluation Center, University of Michigan, 1957.

95. Sabin AB, et al. Live orally given poliovirus vaccine. Effects of rapid mass immunization on populations under condition of massive enteric infection with other viruses. *JAMA.* 1960;173:1521–1526.

96. Lum LCS, Wong KT, Lam SK, et al. Fatal enterovirus 71 encephalomyelitis. *J Pediatr.* 1998;133:795–798.

Chapter 15

97. Ho M, Chang FY, Yin HC, Ben RJ, Chang LY, Chen PY, et al. Antibiotic usage in community-acquired infections in hospitals in Taiwan. *J Formos Med Assoc.* 2002;101:34–42.

98. Ho M, McDonald LC, Lauderdale TL, Yeh LL, Chen PC, Shiau YR, and participating hospitals. Surveillance of antibiotic resistance in Taiwan, 1998. *J Microbiol Immunol Infect.* 1999;32:239–249.

99. A. Fleming in The New York Times, p. 21, June 26, 1945.

100. Cars O, Molstad S, Melander A. Variation in antibiotic use in the European Union. *Lancet.* 2001;357:1851–1853.

101. Chang MT. The abuse of antimicrobials endangers the health of Taiwan's people (in Chinese). *The Control Yuan.* Taiwan, ROC, 1999.

102. Sappala H, Klaukkat, Vuopio-Varkila J, et al. The effect of changes in the consumption of macrolide antibiotics on erythromic resistance in group A streptococci in Finland. *N Engl J Med.* 1997;337:441–446.

103. McDonald LC, Yu HT, Yin HC, Hsiung CA, Ho M, and the Antibiotic Use Working Group. The use and abuse of surgical antibiotic prophylaxis in Taiwan hospitals. *J of Formosa Med Soc.* 2001;100:5–13.

104. Ho M, Hsiung CA, Yu HT, Chi CL, Chang HJ. Changes before and after a policy to restrict antimicrobial usage in upper respiratory infections in Taiwan. *Int J Antimicrob Agents.* 2004;23:438–445.

105. Hsiung CA, private communication.
106. Hsueh PR, Shyr JM, Wu JJ. Decreased erythromycin use after antimicrobial restriction for undocumented bacterial upper respiratory infections significantly reduced erythromycin resistance in *Streptococcus pyogenes* in Taiwan. *CID*. 2005;40:903–904.
107. Lauderdale TL, private communication.
108. McDonald LC, Chen MT, Lauderdale TL, Ho M. The use of antibiotics critical to human medicine in food-producing animals in Taiwan. *J Microbial Immunol Infect*. 2001;34:97–102.
109. Aerestrup FM, Bager F, Jensen NE, et al. Surveillance of antimicrobial resistance in bacteria isolated from food animals to antimicrobial growth promoters and related therapeutic agents in Denmark. *APMIS*. 1998;106:606–622.

Chapter 18

110. Williams T. *The Complete Illustrated Guide to Chinese Medicine*. New York: Barnes and Noble; 1996. (256 pages.)
111. Zhang Zhongjing. *Treatise on febrile diseases caused by cold*. Translated by Luo Xiwen. Beijing: New World Press.
112. Erslev AJ. The discovery of erythropoietin. *ASAIO Journal*. 1993;39:89–92.
113. Hsieh KH. Evaluation of efficacy of traditional Chinese medicines in the treatment of childhood bronchial asthma: clinical trial, immunological tests and animal study. *Ped Allergy and Immunol*. 1996;7:130–140.
114. Sagan C. *The Demon-Haunted World. Science as a Candle in the Dark*. New York: Ballantine Books; 1996.
115. Medical Research Council. Streptomycin in Tuberculosis Trials Committee. Streptomycin treatment of pulmonary tuberculosis. *British MJ*. 1948;2:769–783.
116. Doll R, Hill AB. Smoking and carcinoma of the lung. Preliminary report. *British MJ*. 1950;2:739–748.
117. Chen KK, Schmidt CF. The action of ephedrine, an alkaloid of Ma Huang. *Proc Soc Exp Biol Med*. 1924;21:351–354.
118. Nortier JL, Martinez MC, Schmeiser NH, et al. Urothelial carcinoma associated with the use of a Chinese herb (Aristolochia fangchi). *New Eng J Med*. 2000;342:1686–1692.

Chapter 19

119. Smith H. *The World's Religions, Our Great Wisdom Traditions*. Harper SanFrancisco, A Division of Harper Collins Publishers; 1991.
120. Wilber K. *Sex, Ecology, Spirituality. The Spirit of Evolution*. 2nd ed. Boston: Shambhala Publications, Inc.; 1995, 2000.
121. Dalai Lama. *Advice on Dying and Living a Better Life*. New York: Atria Books; 2002.

SIGNIFICANT PAPERS 1958–2004
(OUT OF 285)

1. Ho, M. and Kass, E.H.: Protective effect of components of normal blood against the lethal action of endotoxin. *J. Lab. Clin. Med.* 51:297–311, 1958.
2. Ho, M. and Enders, J.F.: An inhibitor of viral activity appearing in infected cell cultures. *Proc. Natl. Acad. Sci. USA* 45:385–389, 1959.
3. Ho, M.: Interferons. *N. Engl. J. Med.* 266:1–15, 1962.
4. Ho, M.: Inhibition of the infectivity of poliovirus ribonucleic acid by an interferon. *Proc. Soc. Exp. Biol. Medi.* 107:639–644, 1961.
5. Ho, M.: Interferon-like viral inhibitor in rabbits after intravenous administration of endotoxin. *Science.* 146:1471–1472, 1962.
6. Ho, M., Kono, Y. and Breinig, M.K.: Tolerance to the induction of interferons by endotoxin or virus. *Proc. Soc. Exp. Biol. Med.* 119:1227–1232, 1965.
7. Ho, M. and Kono, Y.: Effect of actinomycin D on virus and endotoxin-induced interferon-like inhibitors in rabbits. *Proc. Natl. Acad. Sci. USA* 53:220–224, 1965.
8. Ho, M. and Ke, Y.H.: The mechanisms of stimulation of interferon production by a complexed polyribonucleotide. *Virology.* 40:693–702, 1970.
9. Tan, Y.H., Armstrong, J.A., Ke, Y.H. and Ho, M.: Regulation of cellular interferon production: Enhancement by antimetabolites. *Proc. Natl. Acad. Sci. USA* 67:464–471, 1970.
10. Ho, M., Tan, Y.H. and Armstrong, J.A.: Accentuation of production of human interferon by metabolic inhibitors. *Proc. Sco. Exp. Biol. Med.* 139:259–262, 1972.
11. Ho, M., Ke, Y.H. and Armstrong, J.A.: Mechanisms of interferon induction by endotoxin. *J. Infect. Dis.* 128:220–227, 1973.
12. Wu, B.C., Dowling, J.N., Armstrong, J.A. and Ho, M. Enhancement of mouse cytomegaloviurs infection during host versus graft reaction. *Science* 190:56–58, 1975.

13. Ho, M., Suwansirikul, S., Dowling, J.N., Youngblood, L.A. and Armstrong, J.A.: The transplanted kidney as a source of cytomegalovirus infection. *N. Engl. J. Med.* 293:1109–1112, 1975.

14. Ho, Monto: Virus infections after transplantation in man. *Arch. Virol.* 55:1–24, 1977.

15. Pazin, G.J., Armstrong, J.A., Lam, M.T., Tarr, G.C., Jannetta, P.J. and Ho, M.: Prevention of reactivated herpes simplex infection by human leukocyte interferon after operation on the trigeminal root. *N. Engl. J. Med.* 301:225–230, 1979.

16. Ho, M.: Role of specific cytotoxic lymphocytes in cellular immunity against murine cytomegalovirus. *Infect. Immun.* 27:767–776, 1980.

17. Ho, M., Wajszczuk, C.P., Hardy, A., Dummer, J.S., Starzl, T.E., Hakala, T.R. and Bahnson, H.T.: Infections in kidney, heart and liver transplant recipients on cyclosporine. *Trans. Proc.* 15:2768–2772, 1983.

18. Ho, M., Pazin, G.J., Armstrong, J.A., Haverkos, H.S., Dummer, J.S., Janetta, P.J.: The paradoxical effects of interferon on reactivation of oral herpes simplex infection after microvascular decompression for trigeminal neuralgia. *J. Infect. Dis.* 150:867–872, 1984.

19. Ho, M., Miller, G., Atchison, R.W., Breinig, M.K., Dummer, J.S., Andiman, W., Starzl, T.E., Eastman, R., Griffith, B.P., Hardesty, R.L., Bahnson, H.T., Hakala, T.R. and Rosenthal, J.T.: Epstein-Barr virus infections and DNA hybridization studies in post-transplantation lymphoma and lymphoproliferative lesions: Role of primary infection. *J. Inf. Dis.* 152:876–886, 1985.

20. Ho, M.: An early interferon: "Viral inhibitory factor". *J. Interferon Res.* 7:455–458, 1987.

21. Ho, M., Jaffe, R., Miller, G., Breinig, M.K., Dummer, J.S., Makowka, L., Atchison, R.W., Karrer, F., Nalesnik, M.A. and Starzl, T.E.: The frequency of EBV infection and associated lymphoproliferative syndrome after transplantation and its manifestations in children. *Transplantation* 45:719–727, 1988.

22. Cen, H., Breinig, M.C., Atchison, R.W., Ho, M. and McKnight, J.L.C.: Epstein-Barr virus transmission via the donor organs in solid organ transplantation: PCR and RFLP Analysis of IR2, IR3 and IR4. *J. of Virol.* 65:976–980, 1991.

23. Ho, M.: *Cytomegalovirus: Biology and Infections*, Second Edition, Plenum Medical, NY; 1991. (440 pages).

24. O'Marro, S.D., Armstrong, J.A., Asuncion, C., Gueverra, L. and Ho, M.: The effect of combinations of ampligen and zidovudine or dideoxyinosine against human immunodeficiency viruses in vitro. *Antivir. Res.* 17:169–177, 1992.

25. Ho, M., Armstrong, J., McMahon, D., Pazin, G., Huang, X., Rinaldo, C., Whiteside, T., Tripoli, C., Levine, G., Moody, D., Okarma, T., Elder, E., Gupta, P., Tauxe, N., Torpey, D. and Herberman, R.: A phase 1 study of adoptive transfer of autologous CD8+ T lymphocytes in patients with acquired immunodeficiency syndrome (AIDS)-related complex or AIDS. *Blood* 81:2093–2101, 1993.

26. Ho, M.: Risk factors and pathogenesis of posttransplant lymphoproliferative disorders. *Transplantation Proceedings*. 27:38–40, 1995.

27. Ho, M.: Current outlook of infectious diseases in Taiwan, *J Microbiol Immunol Infect*. 31:73–83, 1998.

28. Ho, M., Chen, E.R., Hsu, K.H., Twu, S.J., Chen, K.T., Tsai, S.F., Wang, J.R. and Shih, S.R. The Enterovirus Type 71 epidemic of Taiwan, 1998. *N. Engl. J. Med.* 341:929–935, 1999.

29. Ho, M., McDonald, L.C., Lauderdale, T.L., Yeh, L.L., Chen, P.C., Shiau, Y.R. and participating hospitals.: Surveillance of antibiotic resistance in Taiwan, 1998. *J. Microbiol. Immunol. Infect*. 32:239–249, 1999.

30. Ho, M.: Taiwan seeks to solve its resistance problems. *Science*. 291(5513): 2550–2551, 2001.

31. McDonald, L.C., Yu, H.T., Yin, H.C., Hsiung, C.A., Ho, M. and the Antibiotic Use Working Group.: The use and abuse of surgical antibiotic prophylaxis in Taiwan hospitals. *J. Formosa Med. Soc. J. Formos Med. Assoc*. 100(1):5–13, 2001.

32. McDonald, L.C., Chen, F.J., Lo, H.-J., Yin, H.C., Lu, P.L., Huang, C.H., Chen, P.C., Lauderdale, T.L. and Ho, M.: Emergence of reduced susceptibility and resistance to fluoroquinolones in Esherichia coli in Taiwan and contributions of distinct selective pressures. *Antimicrob Agents and Chem*. 45:3084–3091, 2001.

33. McDonald, L.C., Chen, M.T.: Lauderdale, T.L., Ho, M.: The use of antibiotics critical to human medicine in food-producing animals In Taiwan. *J. Microbiol. Immunol. Infect*. 34(2):97–102, 2001.

34. Ho, M., Chang, F.Y., Yin, H.C., Ben, R.J., Chang, L.Y., Chen, P.Y. et al.: Antibiotic usage in community-acquired infections in hospitals in Taiwan. *J. Formos Med. Assoc*. 101(1):34–42, 2002.

35. Ho, M., Hsiung, C.A., Yu, H.T., Chi, C.L., Yin, H.C. and Chang, H.J.: Antimicrobial usage in ambulatory patients with respiratory infections in Taiwan. *J. Formos Med. Assoc*. 103:96–103, 2004.

36. Ho, M., Hsiung, C.A., Yu, T.Z., Chi, C.L. and Chang, H.J.: Changes before and after a policy to restrict antimicrobial usage in upper respiratory infections in Taiwan. *Intern. J. Microb. Agents*. 23:438–445, 2004.

CHRONOLOGY OF THE LIFE OF MONTO HO

1927   March 28, born in Tau-Hwa Lun, Yiyang, Hunan, China.

1932   Mother Hu Gin-lien died.

1935   Went to Ankara, Turkey with Father, Ho Feng-Shan.

1937   Went to Vienna, Austria, where Father transferred.

1938   Germany invades Czechoslovakia, went with stepmother Grace to Brooklyn, N.Y. to avoid war, returned to Europe in 1939, stayed with Austrian family in Berlin.

1939   September 3, beginning of World War II.

1940   April, entire family settled in Brooklyn, New York. Attended Brooklyn Technical High School.

1941   Went back to China with Father, studied at Ling-Ying Middle School in Hong Kong. Three months later, in December, Japan attacks Pearl Harbor and occupies Hong Kong.

1942   Fled Japanese occupation through "no man's land" in Guangdong, graduated from Ge-Lien (Christian Associated) junior middle school in June. Transferred to Sin-Yi Middle School in Tau-Hwa Lun, Hunan.

1943   Transferred from Sin-Yi to Nankai Middle School, Chungking, where Father had now returned from the U.S.

1945   Graduated from Nankai Senior Middle School, accepted as a major in chemistry, Southwest Associated University, Kunming, Yunnan.

1946   World War II ends, Southwest Associated University dissolved, returned to one of its previous components, Tsinghua University in Beijing. Changed major from chemistry to political science.

1947 Transferred as a junior to Harvard College, Cambridge, Massachusetts. Majored in "philosophy and government".

1949 Graduated from Harvard College, A.B. *magna cum laude*. Admitted to Graduate School, Stanford University, Department of Political Science.

1950 Admitted to Medical School, Stanford University.

1951 Sister Manli born in Cairo, Egypt.

1952 Married Carol Tsu, transferred as junior to Harvard Medical School.

1954 Graduated from Harvard Medical School (M.D.). Residency at Boston City Hospital.

1956 Daughter Bettie Pei-wen born.
Research fellow at Thorndike Memorial Lab, Boston City Hospital, with Edward J. Kass.

1957 Research fellow at Children's Hospital with John F. Enders. Studies on interferon.

1959 Appointed assistant professor, Graduate School of Public Health and School of Medicine, University of Pittsburgh.
Son John Chia-wen born.

1962 Elected "Young Turk", member, American Society of Clinical Investigation, Inc.

1965 Promoted to full professor with tenure.

1971 Appointed Chief, Division of Infectious Diseases, Department of Medicine, School of Medicine.

1974 Appointed Chair, Department of Infectious Diseases and Microbiology, Graduate School of Public Health, University of Pittsburgh. Elected to Association of American Physicians.

1978 Elected to Academia Sinica, Taiwan.
Recipient, Macy Senior Medical Scholar Award.

1992 Named among "Best Doctors of America".
Elected Fellow, American Association for the Advancement of Science.

1997 Retired from the University of Pittsburgh.

1997 Appointed Distinguished Investigator, National Health Reasearch Institutes (NHRI), Taiwan.

2001 Recipient of Distinguished Research Award, NHRI, Taiwan
Recipient of Distinguished Service Medal, First Class, Department of Health, Taiwan.

2002 Publication of *My Education and My Path in Medicine* (in Chinese), New News Publishing Company, Taiwan. Retired from NHRI, Taiwan.

INDEX